Software-Assisted Tailoring of Process Descriptions

Jan Ittner

Software-Assisted Tailoring of Process Descriptions

VDM Verlag Dr. Müller

Imprint

Bibliographic information by the German National Library: The German National Library lists this publication at the German National Bibliography; detailed bibliographic information is available on the Internet at http://dnb.d-nb.de.
  Any brand names and product names mentioned in this book are subject to trademark, brand or patent protection and are trademarks or registered trademarks of their respective holders. The use of brand names, product names, common names, trade names, product descriptions etc. even without a particular marking in this works is in no way to be construed to mean that such names may be regarded as unrestricted in respect of trademark and brand protection legislation and could thus be used by anyone.

Cover image: www.purestockx.com

Publisher:
VDM Verlag Dr. Müller Aktiengesellschaft & Co. KG , Dudweiler Landstr. 125 a, 66123 Saarbrücken, Germany,
Phone +49 681 9100-698, Fax +49 681 9100-988,
Email: info@vdm-verlag.de

Zugl.: Erlangen, FAU, Diss., 2006.

Produced in USA and UK by:
Lightning Source Inc., La Vergne, Tennessee, USA
Lightning Source UK Ltd., Milton Keynes, UK
BookSurge LLC, 5341 Dorchester Road, Suite 16, North Charleston, SC 29418, USA

ISBN: 978-3-8364-6710-0

Abstract

Agile software development does not lessen the necessity of consciously thinking about processes. On the contrary, agile paradigms need to be embedded in a broader perspective on *appropriate* processes: The degree of flexibility a process allows for should always be adjusted to the needs of the current project. *Process tailoring* is an effective means for carrying out these adjustments. However, process tailoring is a difficult task that requires both experience and good knowledge about the current project. In order to facilitate process tailoring, and to promote it being put to use more commonly, we put forward a formal tailoring framework as the basis for a software-based *tailoring support system*.

Our tailoring framework allows for expressing formal *tailoring guidelines*, following a three-part structure comprising available *tailoring options*, a *tailoring universe* that defines metrics about the context in which tailoring takes place, and *tailoring hypotheses* that express, in the form of logic propositions, the dependencies and limitations governing choices of individual options in consideration of the context. Tailoring options represent binary choices about items in a process description; the act of tailoring consists of deciding for each item whether to keep it in the process description, or whether to exclude it. Authors of tailoring guidelines can express *tailoring rules* independently of single tailoring options; a special mechanism automatically transforms them into equivalent, option-specific tailoring hypotheses. Given a process model supplemented with such tailoring guidelines, a tailoring system based on our framework can be used to tailor a concrete process description to the requirements of a specific context.

Tailoring takes place in three steps. First, the user of the tailoring system characterises the current *tailoring context* by providing measurements and estimates for the metrics set out in the tailoring universe. Then he invokes the system's optimisation algorithm to obtain the optimal *tailoring configuration*. The system ranks competing tailoring configurations based on fuzzy-logic valuations of the tailoring hypotheses, considering both the tailoring context and the tailoring decisions. Last, the user reviews the tailoring configuration recommended by the system and either accepts it, or revises some of the tailoring decisions and then lets the system re-optimise the remaining ones. Due to a justification mechanism, the tailoring system can reveal for each tailoring decision which aspects of the tailoring context and which other tailoring decisions support or speak against it.

We close with a sample application of our framework in the context of the *ReqMan* project, and discuss related approaches to tailoring assistance.

Part of the material presented in this publication has been published in:

ITTNER, Jan: *Appropriate Processes: Tailoring Agile Processes*. In *Proceedings of the 3rd World Congress for Software Quality*, September 26–30, 2005, Munich, Germany

Contents

Contents

1 Introduction

1.1 Tailoring Appropriate Processes

The goal of software engineering, as with any other engineering discipline, is to produce products of high quality using as few resources as possible. Typical quality requirements in software engineering are reliability, efficiency, maintainability, usability, and many others. In the late 1960s, software engineers realised that such quality requirements could only be guaranteed by "following a disciplined flow of activities" [CG98]. This insight led to the definition of *software life-cycles* [Roy70], and today is mostly subsumed under the notion of *software processes*.

Early software processes were inspired by paradigms of classical industrial production. Over the years it turned out that these processes could not adequately solve some quality problems that were due to the unique, intangible nature of software. This insight led to a radical paradigm shift around the turn of the millennium, initiated by the Agile Movement [AgA]. The term *agility* soon became synonym with highly adapted, lean and flexible processes that differed fundamentally from their rigid, formal, and strictly plan-driven predecessors.

In the meantime, the Agile Movement has long ended its rebel days; agile ideas have steadily diffused into the every-day business of software development, especially in small or medium-sized teams [Cop01, Miš05]. Without doubt agile software development has brought many benefits in terms of flexibility and individuality both on the personal and on the organisational level. Still, as with all innovations, it would be a mistake to take agile methods as a general replacement for traditional software engineering methods [SR03]. As we are now going to argue, software projects will be the most successful if they adhere to neither of the two extreme views of processes, but instead integrate aspects of both rigid and agile approaches, balancing them off against each other with regard to the project's specific characteristics.

1.1.1 We Cannot Get Around Processes

The first of the four central theses of the Agile Manifesto, put forward by the Agile Alliance, states:

"[We value] individuals and interactions over processes and tools." [AgM]

This statement has led to misunderstandings about—or at least to a narrowed view of—the term *process*. It suggests that processes are an aspect of software development that can be weighed against other aspects, such as "individuals" and "interactions." The erroneous implication is that processes are an optional constituent of a software development project that can be included to a greater or lesser extent [Gla01].

The on-line encyclopedia *Wikipedia* defines a proccess as

> "a naturally occurring or designed sequence of operations or events, possibly taking up time, space, expertise or other resource, which produces some outcome." [Wikd]

Here we come upon a different perspective: Processes are an inherent aspect of all kinds of outcome-oriented actions or events, including software development projects. The question with regard to software development, including agile software development, is therefore not about allowing for *less* or *more* process in a project, but whether or not to be *aware* of the processes underlying a project. Not to reflect on processes is a missed chance for improvements. Having a clear process is also one of the central requirements imposed by standards for software process assessment and improvement; CMMI level 3 demands *established* processes [SEI02a], and similarly SPiCE level 3 requires a *defined* process [ISO98b].[1]

It does not come as a surprise, then, that also from the Agile Movement have originated several proposals for software development processes, most notably *eXtreme Programming* (XP) [Bec01], but also other approaches such as *Crystal* [Coc02] or *Scrum* [SB02].

1.1.2 Appropriate Processes

Agility is usually associated with practices that allow for more flexibility and less overhead in day-to-day project work. So far the focus of the agile community has been mainly on techniques that facilitate the actual production of software. Yet most proponents of agile processes offer only descriptions of monolithic processes that do not allow for adaptation to the specific requirements of individual software engineering projects [SR03].

One of the fundamental ideas of the Agile Movement is the notion that software is a living organism that should be allowed to change as flexibly as its requirements will. Short release cycles aim at ensuring that discrepancies between the software's functionality and its real-world requirements are detected as early as possible. *Refactoring*, an agile approach to restructuring source code according to established patterns and principles, has evolved as a discipline in its own right [Fow99]. To ensure a high degree of agility, the same flexibility should also pertain to the development process driving the project. The

---

[1]We will treat CMMI and SPiCE in greater depth in Chapter 5.

agile perspective should not be limited to the design of work methods (the *how*), but should also affect the *choice* of methods and means (the *what*), thereby flexibly adjusting the "weight" of the methods according to the changing needs of the living project.

The Appropriate Process Movement [APM] has coined the term *Appropriate Process* for processes that are ideally adjusted to a given situation. Instead of looking at heavyweight, rigourous processes and lightweight, agile processes as two irreconcilable opposites, the movement views these two concepts as the extremes of a scale upon which every process will have to be placed differently depending on the situation:

"A process should be as agile as possible, and as robust as necessary." [Oldo3]

The goal, thus, is the departure from "one size fits all" processes, and a turn towards flexible models of processes that can be adapted to the needs of the current project. This kind of adaptation is known as *process tailoring* especially in the realm of traditional, heavyweight and highly formalised process models such as the Rational Unified Process, or process reference models such as CMMI or SPiCE. Nevertheless, tailoring should be equally helpful in environments that do not throughout—if at all—impose strict formal demands on projects. Tailoring here can still help adjust the ideal weight of development processes, adhering to the agile principle of "as little as possible, as much as needed."

1.1.3 Tailoring

Tailoring is a configuration task. As a tailor fits a suit to its wearer, tailoring a process means making it fit the current project. No tailor re-invents a new pattern for every suit he makes, instead he recurs to existing patterns which he adapts according to a number of standardised measurements he takes from his customer. The approach to process tailoring is similar; tailoring here consists in adapting a pre-defined, generic process model to a given purpose and situation, mainly by choosing and adjusting components of the process description [HVo3]. However, among the existing approaches to process tailoring we have examined, we have found only few approaches that include up-front measurements or estimates, and, if they do, then only to a very limited extent.[2]

Tailoring is a difficult task, considering the complexity of many standard process models. The task of tailoring gets even more overwhelming when there is no sufficient experience with a complex process model, and no support available to compensate for that. In many such situations, access to generalised tailoring knowledge makes tailoring easier [Xuo5]. Tailoring is further facilitated if metrics are available for assessing the current project goals and characteristics [LR93].

---

[2]See Chapter 5 for a survey of present approaches to tailoring.

1.2 Theses

In our survey of existing approaches to process tailoring, we did not find an approach that supplies a unified, generic solution for the problem of tailoring: According to our above observations, such a solution should provide effective support for tailoring appropriate processes, based on both experience knowledge and characteristics of the current project. In this work, we want to make a contribution to filling this gap. By supplying both the formal foundations and a concrete software implementation of a generic tailoring framework, we want to lay the foundations for software-assisted tailoring. To get clearer about what can be expected of an effective contribution to software-assisted tailoring, we put forward six theses.

The principal goal and intended contribution of our work is to facilitate process tailoring by means of an appropriate software tool. Our first thesis states that this goal is attainable:

Thesis 1 (effective tool support for process tailoring) *A software tool can provide effective assistance for process tailoring.*

The complexity of tailoring stems from the many conditions and dependencies that need to be considered. Providing support for tailoring requires making these conditions and dependencies explicit in the form of tailoring guidelines, which represent generalised tailoring knowledge and serve to make tailoring recommendations in consideration of the current situation. If such guidelines are to be applied by means of a software tool (Thesis 1), they need to be expressed formally, in an adequate language:

Thesis 2 (adequate and unambiguous formal language for tailoring guidelines) *Tailoring guidelines can be expressed adequately and unambiguously in a formal language.*

There are many different notions of formal language in fields such as mathematics or linguistics [Wikb]. The core property of formal languages we are concerned with in Thesis 2 is unambiguousness so that a software tool is able to interpret expressions in that language. By adequacy we mean that the expressiveness of the formal language is well adapted to the domain in which tailoring guidelines are to be applied.

In order to provide effective tailoring assistance, we further claim that a tailoring system can be designed such that it can weigh competing tailoring configurations against each other (Thesis 3) in order to find the optimal tailoring configuration in terms of the tailoring guidelines (Thesis 4).

Thesis 3 (ordering tailoring configurations by rating) *Formal tailoring guidelines can be used to calculate a rating for every possible tailoring configuration that induces a total* *ordering relation on tailoring configurations, allowing them to be weighed against each other in terms of their appropriateness in a given context.*

Thesis 4 (efficient detection of the optimal tailoring configuration) *It is possible to devise an efficient algorithm that finds the optimal tailoring configuration with regard to its rating.*

The success of a software tool for tailoring depends on its acceptance, which in turn requires easy handling and maintainability. This must hold with regard both to authors of tailoring guidelines, and to users applying the tailoring guidelines through the tailoring software. For authors of tailoring guidelines, the tailoring system must be maintainable:

Thesis 5 (maintainability and scaleability) *It is possible to devise a tailoring system that ensures the maintainability of tailoring guidelines even if they grow large and complex.*

Users seeking tailoring assistance must be able to understand the reasons for the tailoring suggestions put forward by the software:

Thesis 6 (transparent ratings of tailoring configurations) *It is possible to present a tailoring configuration along with its rating in such a way that the reasons for its rating are comprehensible without knowledge of the underlying tailoring rules.*

1.3 Structure of this Work

In this work, we put forward a formal framework for expressing and applying tailoring guidelines, with the intention of implementing that framework as a software library, thus providing the foundation for a *tailoring support system* (TSS) based on a graphical user interface.

In Chapter 2 we lay the foundations for such a tailoring framework and supply formal concepts of options available for tailoring, characteristics of the context in which tailoring takes place, and conditions that serve to assess tailoring decisions on the grounds both of the given context, and of dependencies with other tailoring decisions. Based on these concepts, we also introduce a rating scheme that makes it possible to weigh different tailoring configurations against each other. We then supply an optimisation algorithm for finding the optimal tailoring configuration with respect to the tailoring guidelines.

A TSS will have two principal kinds of users—the process modeler who sets forth the tailoring guidelines, and the process tailorer who uses the TSS to assist him in tailoring a concrete process. In Chapter 3 we propose extensions of our framework that help either group to perform their tasks more easily: We will discuss how the process modeler can be enabled to express tailoring rules in more flexible ways than allowed for by the core tailoring framework. To help the process tailorer understand tailoring recommendations and ratings put forward by the TSS, we also develop a justification mechanism that reveals all relevant facts that have contributed to a particular rating.

To verify the applicability of our approach, and to give an expression of its workings, we discuss a practical application of the pilot implementation of our framework in Chapter 4. The example is taken from a process framework for requirements management, developed in the context of the *ReqMan* project.

In Chapter 5 we give a review of related approaches to tailoring assistance. We examine the V-Model XT in particular detail, because it is currently the only established process standard including rudimentary software-based tailoring guidance.

We close by assessing the results of our work in Chapter 6, and point out some remaining challenges we consider worthwhile for future research.

2 A Tailoring Framework

Effective support for tailoring requires clear and concise tailoring guidelines. If tailoring is to be supported by a software tool, these guidelines must have clearly defined, unambiguous semantics such that they can be interpreted by the software. We therefore now propose a formal framework for tailoring.

We start off by giving working definitions of tailoring in general, and of process tailoring in particular, in Section 2.1. Then, in Section 2.2, we establish the foundation of a *tailoring support system* (TSS) by formally defining all relevant aspects of tailoring. We use fuzzy interval logic (Section 2.3) to set up a method for rating competing tailoring configurations in Section 2.4, and provide an algorithm in Section 2.5 that determines the tailoring configuration with the best rating with regard to given tailoring guidelines and a characterisation of the current context.

Appendix A summarises our conventions for mathematical notation.

2.1 Working Definitions

Our aim in this section is to explain our notions of some key concepts related to tailoring. Many definitions have already been put forward for these concepts elsewhere in literature. We do not intend to add on to these definitions in general, but instead provide our own working definitions in order to achieve maximum clarity within the scope of this work.

2.1.1 Tailoring

Ginsberg and Quinn of the *Software Engineering Institute* (SEI) [SEI] at Carnegie Mellon University give a general definition of tailoring as "the act of adjusting the definitions and/or particularising the terms of a general description to derive a description applicable to an alternate (less general) environment" [GQ94]. For our purposes, we restrict ourselves to a simplified variant:

Definition 1 (Tailoring) *Tailoring is the act of adapting a general description to derive a description applicable to a less general environment.*

Due to this definition, tailoring depends on two inputs—a general description, and a given environment—and produces as its output a description adapted to this environment. Consequently, *process tailoring* is the act of adapting a *process description* to the environment of that process. To arrive at a more detailed definition of process tailoring, we thus need a clear definition of processes and process descriptions.

2.1.2 Processes and Process Tailoring

As is to be expected, there is a wide range of definitions for the term *process* in literature, varying in scope, content, and level of abstraction. In the field of software engineering, the software lifecycle standard ISO/IEC 12207 defines a process as "a set of interrelated activities, which transform inputs into outputs" [ISO95]. Recent publications of the SEI refer to *Webster's Dictionary* for a similarly broad definition [SEI02a]. In Section 1.1.1 we have cited Wikipedia's definition of processes:

> "A process is a naturally occurring or designed sequence of operations or events, possibly taking up time, space, expertise or other resources, which produces some outcome." [Wikd]

Another important aspect of processes is contributed by Conradi's definition of a *development process* as

> "all the real-world elements involved in the development and maintenance of a product (i. e., artefacts, production support tools, activities, agents, process support)." [CFF94]

This definition explicitly states one property about processes that is often silently passed over: Processes are part of the real world as opposed to *descriptions* of processes, which are abstract representations of real processes [CFF94]. We therefore define processes as follows:

Definition 2 (Process) *A process is a naturally occurring or designed sequence of real-world operations or events, possibly taking up time, space, expertise or other resources, which produces some outcome.*

Whenever processes are expected to be reproducible, i. e., repeatable, it is necessary to map out a *process description*. For our purposes, we want to come up with a notion of process descriptions that is as generic as possible, yet is specific enough to allow us to provide a practical definition of process tailoring. We therefore propose the following definition, inspired by the discussion of process elements in [HMVV04]:

Definition 3 (Process Description) *A process description is a collection of process elements. Every process element has a description and may define relations with other process elements.*

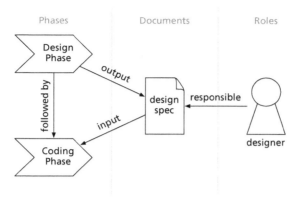

Figure 2.1: Graphical representation of an excerpt from a process description according to Definition 3

Examples for process elements are documents, roles, or phases. Relations between process elements might include a sequential ordering of phases, or responsibilities of roles for documents. Figure 2.1 depicts a sample excerpt from a process description according to Definition 3.

Process descriptions can be helpful in several ways. As *teaching and learning aids*, they serve to provide a first overview to new participants in a project, without necessarily limiting the scope of adaptations to individual preferences or requirements. To those already familiar with the process, they serve as a means of *orientation* rather than a rigid set of instructions, provided sufficient personal and professional maturity of the process participants. Process descriptions also support *communication* between process participants by providing a common point of reference for terms and procedures. Finally, a process description can serve to determine appropriate *tools* to support the process.

Abstract, general process descriptions are often referred to as *process models* [Lon93, FH93]. We provide an according definition:

Definition 4 (Process Model) *A process model is an abstract process description from which concrete process descriptions can be derived for specific environments.*

Now we have all necessary components for a definition of process tailoring:

Definition 5 (Process Tailoring) *Process tailoring is the act of adapting a process model to derive a process description applicable to a given environment.*

9

In the next section, we go on to particularise the nature of the adaptations carried out in the course of tailoring by putting forward a formal tailoring model.

2.2 A Formal Model of Tailoring

So far, we have provided working definitions for tailoring, processes, process descriptions, process models, and process tailoring. Before we proceed to define formal equivalents of these concepts, we must first understand the activities that are usually carried out during process tailoring. According to [HV03], tailoring is performed in four steps:

1. Choose a *process life cycle*. Depending on the size and scope of a project, a simple waterfall model might be sufficient, or an iterative process model is necessary.

2. Choose required *process elements* such as methods, roles, and documents, that constitute the process description. Every process element has a description and may define dependencies with other process elements.

3. Adapt the selected process elements to the context of the project: Checklists, method descriptions, and document outlines can be shortened down in accordance with the complexity of the project.

4. Assign team members to the roles required by the project according to their abilities and preferences.

Step 1 is about the choice of a process life cycle. Within our conceptual framework, we can think of every life cycle as being represented by an associated process model. The choice of process life cycle will in most cases not even be up for decision: In practice, most organisations only use a single process model for a field of activity. Therefore, we will not focus on providing assistance for Step 1.

Steps 2 and 3 can be regarded straightforwardly as choices about a set of binary options: The process model specifies process elements, each of which can be included in the process description being tailored, or can be excluded from it. Likewise, particular constituents of process elements can either be maintained, or deleted.

Step 4 is beyond typical notions of tailoring such as those that have contributed to our concept of tailoring in Section 2.1. Therefore, we do not consider role assignments as an aspect of tailoring, but rather of resource planning. However, tailoring and resource planning are closely interrelated, and there is no sharp boundary between the two. By determining activities, tools, and responsibilities for a project, we already anticipate which resources are going to be required. Conversely, limitations on resources will affect tailoring decisions. Resource planning is a broad field in its own right, and as such it is beyond the scope of our approach to tailoring. Yet we will outline in Section 2.4.6 that

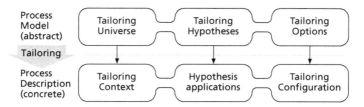

Figure 2.2: Overview of our tailoring model

our tailoring model is well suited to incorporate considerations about resource planning even in the absence of precise data about resource allocations.

To sum up, our formal approach to tailoring will have to take care of steps 2 and 3 discussed above. We have observed that both steps consist in taking binary decisions. Consequently, we define *tailoring options* as the central element of the tailoring procedure:

Definition 6 (Tailoring Option) *Tailoring options are the elementary, binary alternatives representing the choices available in the act of tailoring.*

Although Definition 6 is motivated by considerations about process tailoring, it is remarkable that it does not refer to any specific properties of processes in itself. By looking at tailoring as making a series of binary decisions, we can now develop a formal model of tailoring independently of the process domain. However, to ensure that we stay on track with our initial aim of tailoring appropriate processes, we will accompany our discussion with examples from the domain of software development processes.

We have established in Section 2.1 that tailoring takes a general description and a given environment as its input and produces a description that is adapted to the environment. To provide assistance with tailoring, we also need a third kind of input: We need guidelines for tailoring. To implement a software-based TSS, we need to provide formal equivalents of all inputs and outputs of tailoring. Our formal tailoring model will therefore include specifications of *tailoring options* to represent the general description, a *tailoring universe* to specify a generic tailoring environment, and *tailoring hypotheses* to represent tailoring guidelines. Given a concrete tailoring environment in the form of a *tailoring context*, a *tailoring algorithm* will apply the tailoring hypotheses to the tailoring context in order to produce, as its output, the optimal *tailoring configuration* with respect to the criteria set forth in the hypotheses. Figure 2.2 gives a schematic overview of our tailoring model.

We will now provide detailed formal definitions for tailoring options and tailoring configurations (Section 2.2.1), tailoring universes (Section 2.2.2), and tailoring contexts (Section 2.2.3). We treat tailoring hypotheses only in Section 2.4 after having introduced fuzzy interval logic in Section 2.3.

2.2.1 Tailoring Options and Tailoring Configurations

Let **O** denote the set of available tailoring options. Then we can define a *tailoring config-uration* as a set of associations between tailoring options and binary tailoring decisions. More formally, the set of all possible tailoring configurations is the family of mappings from options in **O** to Boolean values $\mathbb{B}$:

$$\mathbf{C} = \mathbb{B}^{\mathbf{O}} \tag{2.1}$$

For any tailoring configuration $C \in \mathbf{C}$, we call an option *o chosen* if $C(o)$ is true; other-wise, we call the option *excluded*. Our aim now is to find the *optimal* tailoring configura-tion $C^* \in \mathbf{C}$ where the choice of options is optimally suited to the tailoring context, and where conflicts between the chosen options are minimised.

2.2.2 Tailoring Universe

The *tailoring universe* specifies possible *tailoring contexts* in actual tailoring situations. It defines types of *entities* and their *properties*, *metrics*, and *scales*.

Entities and Properties

The context of a TSS can be expressed as a collection of *entities* that are each characterised by a number of *properties*. The tailoring universe defines possible types for these entities along with applicable properties. A central entity type in most cases will be *project*, with properties such as team size, projected duration or budget; other entity types will mainly represent types of stakeholders, e. g., staff, clients, end users, and their respective organisations.

Metrics

Metrics are numeric measurements of properties [Hin96]. Since tailoring is performed at an early stage of a project, most of the metrics will reflect estimates of future conditions (estimate metrics) or evaluations of prior experience (experience metrics). From the per-spective of the TSS, *measuring* these metrics is equivalent to having a questionnaire filled out by the user of the TSS. The metrics that determine the tailoring context need to be compiled and designed such as to encourage accurate and meaningful answers from the user, and, ultimately, to allow for a tailoring configuration that is in line both with the metrics, and with what a human expert would accept as an appropriate configuration. Viewed from the perspective of classical test theory [LR98], this questionnaire is a *test* that has to fulfil three main criteria:

Objectivity Metrics should be *objective*: The result of the tailoring procedure should be independent of which user is carrying it out. Different users with the same knowledge of the tailoring context should be able to produce the same results. A TSS can sustain objectivity by providing a clear user interface and unambiguous descriptions of the metrics under consideration.

Reliability The *reliability* criterion demands that Metrics be designed such that they can be measured *consistently*. Thus, the metrics should provide as little room for subjective deviations as possible. As an example, subjective assessments on a *best–worst* scale usually do not gain accuracy when carried out on a 10-point scale instead of a 5-point scale, simply because the difference of only one point on the scale becomes intangible. A 5-point scale, on the other hand, can be associated with clear notions of *worst*, *below average*, *average*, *above average*, and *best*. At the heart of the reliability criterion is the goal of avoiding a misleading pseudo-accuracy where in fact relevant data is overlaid by the random noise of subjective interpretations or otherwise inaccurate measurements of metrics. However, the criterion of reliability makes no claim about the *correctness* of the results. Furthermore, no promise is made as to the *relevance* of what they measure, that is, whether they provide any relevant insight.

Validity While the reliability criterion only calls for consistent measurements, the criterion of *validity* demands that the conclusions drawn from the initial measurements—in our case a tailoring configuration—are correct. There are various technical definitions of validity. One of these is *content validity*, which demands that the metrics underlying the measurements are drawn from the domain in question. *Predictive validity* is an empirical approach where the conclusions derived from the formalised measurements are confronted with the predictions of human experts. Reliability is a prerequisite for validity, but, as stated above, the reverse does not apply.

Fulfilling these criteria much depends on carefully designing appropriate metrics, but it also requires a good understanding of the computational and semantic characteristics of the data to be collected. Within the field of test theory the latter is addressed by scale theory.

Scale Types

Metrics are specifications of numerical measurements. The mathematical and empirical properties of numbers depend on what they represent. Stanley Smith Stevens, an experimental psychologist, in 1946 put forward a taxonomy of four types of scales [Ste46] that still remains the most widely used today, also in other disciplines such as software metrics [ED96]. Each of these scale types can be distinguished by the operations they permit for comparing values on the same scale.

Names	arbitrary labels
Grades	ordered labels such as Freshman, Sophomore, Junior, Senior
Ranks	starting from 1, which may represent either the largest or smallest
Counted fractions	bounded by zero and one, e. g., percentages
Counts	non-negative integers
Amounts	non-negative real numbers
Balances	unbounded, positive or negative real numbers

Table 2.1: Scale typology proposed by Mosteller and Tukey

Nominal Scale The only permissible operation is comparison for identity, as with numbers on the shirts of football players. In fact, these numbers can be substituted for by a set of arbitrary symbolic identifiers.

Ordinal Scale In addition allows for relative comparisons, e. g., between street numbers.

Interval Scale Adds the possibility of calculating distances between values, e. g., for points in time or on a route, or temperatures measured in Celsius.

Rational Scale Defines an absolute zero point and thus also provides meaning to ratios of values, e. g., for distances, counts, amounts of money, or temperatures measured in Kelvin (which, in contrast to the Celsius scale, defines an absolute zero point).

Stevens' taxonomy of scale types has been subject to criticism about deficiencies of statistical claims associated with the taxonomy, but also for its incompleteness [VW93]. Our tailoring approach is not based on statistical methods, so we can safely ignore most of these objections; however, we will gain expressiveness by extending and differentiating Stevens' taxonomy. We therefore adopt the scale types proposed by Mosteller and Tukey [MT77, VW93] (Table 2.1). Names, grades, and ranks in most cases correspond to nominal, ordinal, and interval scales, respectively. All other scale types typically relate most closely to Stephen's rational scale, although they impose further limitations such as number type—e. g., integer versus real numbers—and restrictions on range.

Formal Definition

We will now construct a formal basis for tailoring universes. A tailoring universe specifies a set of entity types T. Each entity type is associated with a number of properties in P:

$$properties : T \mapsto \mathcal{P}(P) \qquad (2.2)$$

Example 2.1 *Assuming entity types*

$$T =_{def} \{project, staff, client\}$$

we could define the properties

$$properties(\text{project}) =_{\text{def}} \{\text{budget, duration, team-size}\}$$
$$properties(\text{staff}) =_{\text{def}} \{\text{experience}\}$$
$$properties(\text{client}) =_{\text{def}} \{\text{accessibility}\}$$

Every property can be measured on a scale. Metrics associate properties $p \in \mathbf{P}$ with scales $S \in \mathbf{S}$:

$$metric : \mathbf{P} \mapsto \mathbf{S} \qquad (2.3)$$

Example 2.2 *Property* "budget" *is associated with a scale that measures amounts of money:*

$$metric(\text{budget}) =_{\text{def}} S_{\text{money}}$$

Now all that is left is to provide the means for formally specifying scales. We distinguish different scale types as defined in Table 2.1:

$$scaletype : \mathbf{S} \mapsto \{\text{name, grade, rank, counted-fraction, count, amount, balance}\} \qquad (2.4)$$

Let $\mathbf{S}_{t_1,\ldots,t_n} \in \mathbf{S}$ denote the subset of scales of types $t_{1\ldots n}$, and let $\mathbf{I}$ denote a set of identifiers. Each scale of type 'name' is associated with an unordered subset of $\mathbf{I}$:

$$id : \mathbf{S}_{\text{name}} \mapsto \mathcal{P}(\mathbf{I}) \qquad (2.5)$$

Similarly, each scale of type 'grade' is associated with an ordered list of identifiers:

$$id : \mathbf{S}_{\text{grade}} \mapsto \{\langle I_1, \ldots, I_k \rangle \mid k > 0 \wedge (\forall 1 \le i, j \le k)(I_i, I_j \in \mathbf{I} \wedge i \ne j \Leftrightarrow I_i \ne I_j)\} \qquad (2.6)$$

where we have a total ordering relation $<$ over ordered lists $I = \langle I_1, \ldots, I_k \rangle$:

$$I_i < I_j \Leftrightarrow_{\text{def}} i < j \qquad (2.7)$$

Scales of all other types except ranks may be associated with a unit of measurement in $\mathbf{U}$:

$$unit : \mathbf{S}_{\text{counted-fraction,count,amount,balance}} \mapsto \mathbf{U} \qquad (2.8)$$

Each scale provides a domain of valid measurements:

$$\text{dom}(S \in \mathbf{S}) = \begin{cases} id(S) & \text{if } S \in \mathbf{S}_{\text{name}} \\ \bigcup_{1 \le i \le |id(S)|}\{id(S)_i\} & \text{if } S \in \mathbf{S}_{\text{grade}} \\ \mathbb{N} & \text{if } S \in \mathbf{S}_{\text{rank}} \\ \{x \in \mathbb{R} \mid 0 \le x \le 1\} & \text{if } S \in \mathbf{S}_{\text{counted-fraction}} \\ \mathbb{Z}^* & \text{if } S \in \mathbf{S}_{\text{count}} \\ \{x \in \mathbb{R} \mid x \ge 0\} & \text{if } S \in \mathbf{S}_{\text{amount}} \\ \mathbb{R} & \text{if } S \in \mathbf{S}_{\text{balance}} \end{cases} \qquad (2.9)$$

Example 2.3 *Scale* $S_{money} \in S$ *denotes amounts of money:*

$$scaletype(S_{money}) =_{def} amount$$
$$unit(S_{money}) =_{def} Euro$$

From this follows that

$$\dom(S_{money}) = \{x \in \mathbb{R} \mid x \geq 0\}$$

Example 2.4 *A simple rating scale* S_{rating} *with ratings* {worst, average, best} $\subset I$ *is defined as*

$$scaletype(S_{rating}) =_{def} grade$$
$$id(S_{rating}) =_{def} \langle worst, average, best \rangle$$

with

$$\dom(S_{rating}) = \{worst, average, best\}$$

2.2.3 Tailoring Contexts

Every tailoring context is based on a set E of entities. Every entity is assigned an entity type from a tailoring universe T:

$$e\text{-}type : E \mapsto T \qquad\qquad (2.10)$$

Example 2.5 *We have entities* $E = \{minitool, peter, laura\}$ *and associate them with types from the universe in Example 2.1:*

$$e\text{-}type(minitool) =_{def} project$$
$$e\text{-}type(peter) =_{def} staff$$
$$e\text{-}type(laura) =_{def} staff$$

In the tailoring universe, every entity type is associated with a number of properties (2.2). In a tailoring context, each entity-property pair takes the role of a variable. We define the set of all variables for a set of entities E as:

$$V = \{(e, p) \mid e \in E \land p \in properties(e\text{-}type(e))\} \tag{2.11}$$

Example 2.6 *From Examples 2.1 and 2.5 we get the following variables:*

$$V = \{(\text{minitool, budget}), (\text{minitool, duration}), (\text{minitool, team-size}),$$
$$(\text{peter, experience}), (\text{laura, experience})\}$$

Each variable can be measured by virtue of the metric of the associated property. The common domain of all measurements is the union of the domains of all measurement scales:

$$D = \bigcup_{S \in \mathbf{S}} \mathrm{dom}(S) \tag{2.12}$$

We define measurements in the tailoring context as the mapping

$$measurement : V \mapsto D \tag{2.13a}$$

where, more specifically,

$$measurement(e, p) \in \mathrm{dom}(metric(p)) \cup \{\perp\} \tag{2.13b}$$

Measurements need not be defined for all combinations of entities and properties; to avoid having a partial measurement function we denote undefined measurements by $\perp$.

In practice, instead of determining a single exact value for a measurement, it is often easier to state only the lower and upper bounds of a range of plausible values:

$$\widehat{measurement} : V \mapsto \{[u, v] \mid u, v \in D, u \leq v\} \tag{2.14a}$$

with

$$\widehat{measurement}(e, p) \in \{[u, v] \mid u, v \in \mathrm{dom}(metric(p)), u \leq v\} \cup \{\perp\} \tag{2.14b}$$

Example 2.7 *We can define the budget of project "minitool" in the following way:*

$$\widehat{measurement}(\text{minitool, budget}) =_{\mathrm{def}} [5000, 6000] \in \{[u, v] \mid u, v \in \mathrm{dom}(S_{\mathrm{money}})\}$$

According to Example 2.3 we have

$$unit(metric(\text{budget})) = unit(S_{\mathrm{money}}) = \mathrm{Euro}$$

A tailoring context **K** for tailoring universe **T** can be described by specifying a set of entities **E** and a range measurement mapping:

$$\mathbf{K} = (\mathbf{E}, \widehat{measurement}) \tag{2.15}$$

Example 2.8 *With entities* **E** *from Example 2.5 and the—partially defined—measurement function from Example 2.2.3 we get the tailoring context*

$$\mathbf{K} = (\{\text{minitool, peter, laura}\}, \{(\text{minitool, budget}) \rightarrow [5000, 6000]\})$$

2.2.4 Optimal Tailoring Configurations

In sections 2.2.1 through 2.2.3 we have provided definitions for tailoring contexts, universes, and configurations. To accomplish an automated TSS as outlined in Figure 2.2, we still need to provide a way to formally express tailoring hypotheses.

The tailoring hypotheses should serve two purposes:

1. They should allow the TSS to rate tailoring configurations and thus provide a frame of reference for finding *optimal* tailoring configurations, and

2. they should allow the TSS to provide justifications for recommended tailoring configurations to the user.

Tailoring hypotheses fulfilling these requirements will associate tailoring options with demands about the tailoring context, including dependencies between different tailoring options. Once we can rate tailoring configurations, we need to specify an algorithm that calculates the optimal tailoring configuration.

The following sections discuss the structure of hypotheses, a rating method and optimisation algorithm. In order to attain sufficient expressiveness without compromising transparency to the user, we will express tailoring hypotheses as terms in *fuzzy logic* whose atomic constituents are *fuzzy variables*. Fuzzy variables can be devised to represent intuitive statements in natural language. Provided the logic terms built around these variables are not overly complex, a simple mechanism can transform these terms into representations in everyday language which a user can understand intuitively.

There is another reason for representing hypotheses with fuzzy logic: Most real-world tailoring decisions are not black-and-white. Consider, for example, a tailoring hypothesis that demands a minimum budget of $5000 for a specific tailoring option. It is not reasonable to vote against the option if the actual budget is $4999 and to vote in favour of it with a budget of $5000. Instead of this sharp cut, the more natural approach is to

Figure 2.3: Applicability of a tailoring option based on available budget

have a recommendation gradient within a range of budgets. We might then say that the option is not recommended below a budget of $4500, and that it is highly recommended above a budget of $5500. For budgets in between these boundaries, the recommendation gets stronger with increasing budget (Figure 2.3). Fuzzy logic allows us to express such conditions.

2.3 Fuzzy Interval Logic

Fuzzy logic was originally introduced by Lotfi A. Zadeh in 1965 [Zad65]; since then a rich body of research literature has evolved on the subject. We now give a quick overview of the variant of fuzzy logic which we are going to use for evaluating tailoring contexts. In the following sections, we introduce fuzzy propositional variables (Section 2.3.1), fuzzy formulæ (Section 2.3.2), and valuation functions (Section 2.3.3).

2.3.1 Fuzzy Propositional Variables

A propositional variable a represents a statement that can be true or false. A valuation function assigns it a truth value:

$$v(a) \in \{\text{false}, \text{true}\} \tag{2.16}$$

A *fuzzy propositional variable* (FPV) can also be valuated to *fuzzy truth values* (FTVs) in between the extremes of *true* and *false*. Intuitively such a value can be interpreted as a grade of applicability of the statement in question. The truth scale ranges from 0 (not at all applicable) to 1 (fully applicable).

$$v(a) \in \mathbf{Q} \tag{2.17}$$

with

$$\mathbf{Q} = \{x \in \mathbb{R} \mid 0 \leq x \leq 1\} \tag{2.18}$$

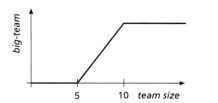 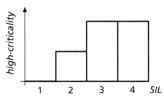

Figure 2.4: fuzzy propositional variables from Example 2.9

Example 2.9 *Properties from the tailoring context can be mapped to FPVs. Let us consider the following two properties:*

property	scale type	domain
team-size	count	$\mathbb{N}$
SIL[1]	name	$\{1, 2, 3, 4\}$

Based on this we can define valuations for the following FPVs (Figure 2.4):

$$v(big\text{-}team) = \begin{cases} 0 & \textit{if } \text{team-size} \leq 5 \\ \frac{\text{team-size}-5}{5} & \textit{if } 5 < \text{team-size} < 10 \\ 1 & \textit{if } \text{team-size} \geq 10 \end{cases}$$

$$v(high\text{-}criticality) = \begin{cases} 0 & \textit{if } \text{SIL} = 1 \\ 0.5 & \textit{if } \text{SIL} = 2 \\ 1 & \textit{if } \text{SIL} \geq 3 \end{cases}$$

2.3.2 Fuzzy Formulæ

We define the set **F** of *fuzzy formulæ* (FFs) as follows:

1. All FPVs are FFs.

2. The constants 'true' and 'false' are FFs.

3. If a is a FF, then also $\neg a$.

4. If a and b are FFs, then also $a \wedge b$ and $a \vee b$

5. There are no other FFs.

---

[1]Safety Integrity Level specified by the IEC 61508 standard, ranging from 1 (low risk) to 4 (high risk)

Example 2.10 *Some valid FFs are:*

$$big\text{-}team \wedge \neg high\text{-}criticality$$
$$high\text{-}criticality$$
$$\neg(big\text{-}team \vee high\text{-}criticality)$$

2.3.3 Valuation Functions

Standard fuzzy logic defines valuation functions that valuate FFs to FTVs. We first introduce this standard valuation function, and then propose an extended valuation model that is also capable of valuating FFs when only partial information is available.

Simple Valuation

The valuation function $v : F \mapsto Q$ is defined as follows:

$$v(p \wedge q) = min(v(p), v(q)) \qquad (2.19a)$$
$$v(p \vee q) = max(v(p), v(q)) \qquad (2.19b)$$
$$v(\neg p) = complement(v(p)) \qquad (2.19c)$$
$$v(\text{false}) = 0 \qquad (2.19d)$$
$$v(\text{true}) = 1 \qquad (2.19e)$$

Where functions *min*, *max*, and *complement* are defined as:

$$min(r, s) = \begin{cases} r & \text{if } r < s \\ s & \text{otherwise} \end{cases} \qquad (2.20a)$$

$$max(r, s) = \begin{cases} r & \text{if } r > s \\ s & \text{otherwise} \end{cases} \qquad (2.20b)$$

$$complement(r) = 1 - r \qquad (2.20c)$$

The rationale behind these definitions is that the FTV of a logical conjunction should not be above the FTV of its weakest subterm, i. e., the subterm with the lowest FTV; similarly, a logical disjunction should not evaluate to a FTV below its stongest subterm. Negation is simply the complement over the scale of FTV.

Handling Partial Information with Fuzzy Interval Logic

We have just given a short account of a standard variant of fuzzy logic. Fuzzy logic goes beyond Boolean logic in that it introduces grades of applicability of statements between the extremes of *false* and *true*, but as with Boolean logic it evaluates to precise, determined results.

With *fuzzy interval logic* (FIL) we now propose an extension to fuzzy logic that will allow us to deal with incomplete or vague information, as will often be the case in tailoring contexts when estimates can not be pinpointed exactly, but only within a range, or even not at all.

Let us therefore assume that the valuations for variables need not be determined exactly. This may become necessary when there is insufficient information to determine the exact result of the valuation function. In that case, we can still provide a meaningful valuation for a term if we allow intervals of possible FTVs instead of exact values. We will refer to these intervals as *fuzzy truth intervals* (FTIs), and write FTIs as $[l, h]$, where l denotes the inclusive lower bound of the interval, and h denotes the inclusive upper bound.

With maximum uncertainty, i.e., total lack of information, we can valuate a FF to the FTI $[0, 1]$, meaning that nothing at all can be said about the applicability of the FF under consideration. On the other hand, maximum certainty is expressed as the point interval $[x, x]$ which represents a maximally exact statement about the applicability of the FF.

The valuation function for FTIs is analogous to (2.19):

$$\hat{v} : \mathbf{F} \mapsto \hat{\mathbf{Q}} \qquad (2.21)$$

with

$$\hat{\mathbf{Q}} = \{[r_l, r_h] \mid 0 \leq r_l \leq r_h \leq 1 \wedge (r_l, r_h) \in \mathbb{R} \times \mathbb{R}\} \qquad (2.22)$$

and

$$\hat{v}(p \wedge q) = \widehat{min}(\hat{v}(p), \hat{v}(q)) \qquad (2.23\text{a})$$
$$\hat{v}(p \vee q) = \widehat{max}(\hat{v}(p), \hat{v}(q)) \qquad (2.23\text{b})$$
$$\hat{v}(\neg p) = \widehat{complement}(\hat{v}(p)) \qquad (2.23\text{c})$$
$$\hat{v}(\text{false}) = [0, 0] \qquad (2.23\text{d})$$
$$\hat{v}(\text{true}) = [1, 1] \qquad (2.23\text{e})$$
$$\hat{v}(\text{unknown}) = [0, 1] \qquad (2.23\text{f})$$

where we have extended the set of valid FFs by the constant "unknown."

This time, we define the underlying functions like this:

$$\widehat{min}([r_l, r_h], [s_l, s_h]) = [min(r_l, s_l), min(r_h, s_h)] \qquad (2.24a)$$

$$\widehat{max}([r_l, r_h], [s_l, s_h]) = [max(r_l, s_l), max(r_h, s_h)] \qquad (2.24b)$$

$$\widehat{complement}([r_l, r_h]) = [1 - r_h, 1 - r_l] \qquad (2.24c)$$

Finally we provide two auxiliary functions to refer to the lower and upper bounds of a FTI:

$$lo([r, s]) = r \qquad (2.25a)$$

$$hi([r, s]) = s \qquad (2.25b)$$

Standard fuzzy logic can be considered a special case of FIL that is restricted to point intervals. One can verify by simple substitution that there is a homomorphism h between the valuation function v (2.19) of Section 2.3.3 and $\hat{v}$ (2.23):

$$h(x \in \mathbb{R}) = [x, x] \qquad (2.26)$$

2.4 Using Hypotheses to Rate Tailoring Configurations

Before we proceed to define tailoring hypotheses based on fuzzy logic, let's take a closer look at the result we want to achieve: Given a tailoring context **K** for a tailoring universe **T**, we want to rate tailoring configurations $C \in \mathbf{C}$:

$$rating : \mathbf{C} \mapsto [0, 1] \equiv [0\%, 100\%] \qquad (2.27)$$

Out of all tailoring configurations, we want to find a tailoring configuration C^* with the highest rating:

$$rating(C^*) = \max_{C \in \mathbf{C}} rating(C) \qquad (2.28)$$

We will define function *rating* based on the ratings for the tailoring decisions making up a tailoring configuration. Recall that a tailoring configuration $C \in \mathbf{C}$ is a mapping $C : \mathbf{O} \mapsto \mathbb{B}$. We call $C(o)$ a tailoring decision because it assigns a Boolean value to option $o \in \mathbf{O}$, indicating whether the option is chosen. We write the rating function for tailoring decisions about options **O** in configuration C as follows:

$$rating_C : \mathbf{O} \mapsto [0, 1] \qquad (2.29)$$

We call options rated 50% or above *recommended*. Options rated below 50% are called *not recommended*. A rating of 0% means that an option is not recommended at all, whereas an option rated 100% implies a strong recommendation. Ratings on the range between 0% and 100% signify recommendations of corresponding strength.

If we express our tailoring hypotheses by means of FFs—as put forward in Section 2.2.4— we can calculate $rating_C$ from valuations of FFs. In Section 2.4.1, we will show how FFs can be used to make statements about the tailoring context. Then, in Section 2.4.2, we will develop a definition for $rating_C$. Finally, we will discuss in Section 2.4.4 how these individual decision ratings can be combined to a definition for rating function *rating* for tailoring configurations.

2.4.1 Fuzzy Formulæ about the Tailoring Context

A FF uses logic operators to combine atomic terms, i.e., constants and FPVs. As discussed in Section 2.3.1, a FPV represents a statement whose applicability can be expressed as a FTV on a scale from 0 *(not applicable)* to 1 *(fully applicable)*. We have already shown in Example 2.9 that measurements from the tailoring context K can be mapped to valuators for FTVs. In the following we present a generic approach for associating FPVs with valuators that incorporate the tailoring context.

Since each FPV can be valuated based on individual rules, we must provide a custom valuator v_a for every FPV a:

$$v(a) =_{\text{def}} v_a \tag{2.30}$$

In the following we discuss some generic classes of valuation functions for FPVs.

Ramps

In standard fuzzy logic, a large family of FPV valuators is derived from *ramps* and ramp-like functions (Figure 2.5). Any property p where $\mathrm{dom}(metric(p)) \subseteq \mathbb{R}$ can be included in a valuator as follows:

$$ramp_{r,s}(x) = \begin{cases} 0 & \text{if } x < r \\ \frac{x-r}{s-r} & \text{if } r \leq x < s \\ 1 & \text{if } x \geq s \end{cases} \qquad \text{where } r \leq s \tag{2.31}$$

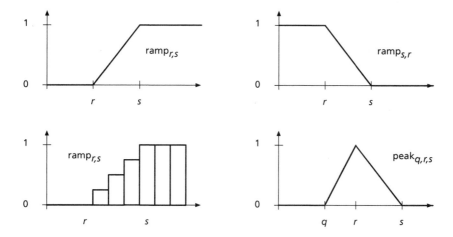

Figure 2.5: Variations of ramp-like mappings for FPVs

Example 2.11 *In parallel with Example 2.9 we can define the valuator[2] for FPV big-team as*

$$v_{big\text{-}team} = ramp_{5,10}(measurement(\text{minitool, team-size})) \qquad (2.32)$$

Note that (2.31) is also defined in the case that $r = s$: We then obtain a "step" function equivalent with the valuation of the Boolean expression $x \geq r$.

For properties that are measured on an interval scale with discrete values, e. g., natural numbers, a ramp function can be defined that replaces the quotient $\frac{x-r}{s-r}$ with an integer division $\lfloor (x - r)/(s - r) \rfloor$.

A ramp can also be defined in the reverse direction, gradually descending from 1 down to 0 between r and s:

$$ramp_{s,r}(x) = 1 - ramp_{r,s}(x) \qquad \text{where } s \geq r \qquad (2.33)$$

Two *ramp* functions can be combined to obtain a *peak* function:

$$peak_{q,r,s}(x) = \min(ramp_{q,r}(x), 1 - ramp_{r,s}) \qquad \text{where } q \leq r \leq s \qquad (2.34)$$

---

[2]Note that this valuator leaves us with no means of dealing with undefined measurements; we will deal with that in Section 2.4.5.

Value Tables

For any property with a finite number of possible measurements, e. g., names or grades, each possible value can be associated with an individual value in a table.

Dependencies between Options

Choosing an option might require that one or more other options be also chosen, or that conflicting options be excluded from the tailoring configuration. We can express tailoring hypotheses that depend on the current tailoring configuration $C \in \mathbf{C}$ if we define FPVs whose valuation depends on a tailoring decision in C:

$$configured(o \in \mathbf{O}) = \begin{cases} 1 & \text{if } C(o) \\ 0 & \text{if } \neg C(o) \end{cases} \qquad (2.35)$$

With this we introduce dependencies between hypotheses and the tailoring configuration: Hypotheses referring to the tailoring configuration can only be valuated if the tailoring configuration is already known. This is a potential drawback when a tailoring configuration is to be constructed incrementally and intermediate ratings need to be calculated in the process. We will show in Section 2.4.5 how this problem can be circumvented so that such hypotheses can be evaluated even when the tailoring configuration is only partially defined.

2.4.2 Tailoring Hypotheses

When assessing decisions about tailoring options, we can distinguish two kinds of considerations: First, we need to decide whether an option is *necessary* in a given context. For reasons of economy, we would expect a TSS to recommend only those options that are really required. Second, we need to ensure that it is *feasible* to implement a specific option.[3] Hence, every tailoring option can be assigned a *necessitating hypothesis* that determines its necessity and a *qualifying hypothesis* that determines its feasibility.

This is in parallel with the concept of necessary and sufficient conditions in the philosophy of logic [Bre03]. The difference in the case of our rules is that, instead of reasoning about facts, we are dealing with modal concepts of obligation and permission—i. e., whether a tailoring option *should* be chosen and whether it *may* be chosen. We will explore this aspect in further detail in Section 3.1.1, where we will introduce a modal operator to

---

[3]For example, it is not enough to realise that an injured driver is in need of first aid; to administer it, someone must be present who has been trained accordingly.

express rules explicitly. For now it suffices to establish that an option $o \in O$ should be rated based on two FFs, denoted as $hyp_n(o)$ and $hyp_q(o)$ with

$$hyp_n : O \mapsto F \tag{2.36a}$$
$$hyp_q : O \mapsto F \tag{2.36b}$$

where the necessitating hypothesis $hyp_n(o)$ valuates to the necessity of option o, and the qualifying condition $hyp_q(o)$ valuates to the feasibility of option o. Let $v_n(o)$ denote the necessity valuation, and $v_q(o)$ denote the feasibility valuation:

$$
\begin{aligned}
v_n(o) &= v(hyp_n(o)) \\
v_q(o) &= v(hyp_q(o))
\end{aligned}
\qquad \text{with } o \in O \tag{2.37}
$$

2.4.3 Rating Tailoring Decisions

We are now left with the task of combining $v_n(o)$ and $v_q(o)$ into a meaningful overall rating for an option $o \in O$. For this we need to take into account whether or not o is chosen or excluded, i.e., whether $C(o)$ is true or false.

Rating Chosen Options

Let us first consider the case that o is chosen. We can establish that the overall rating should not exceed $v_n(o)$ for reasons of economy—an option is only as valuable as it is necessary. Consequently, if $v_q(o) \geq v_n(o)$, the overall rating will be $v_n(o)$. If, however, we have $v_q(o) < v_n(o)$, the overall rating will depend on our readiness and ability to overcome the gap between feasibility and necessity. At the one extreme, which we call *conservative tailoring*, the overall rating is not allowed to exceed $v_q(o)$, that is, we never make recommendations that are stronger than the feasibility of the option in question. Alternatively we can allow the overall rating to exceed $v_q(o)$ by a certain offset. We call this *progressive tailoring*. For chosen options, we thus define the rating as

$$\min(v_n(o), v_q(o) + k) \qquad \text{where } 0 \leq k \leq 1 \tag{2.38}$$

With $k = 0$ we have conservative tailoring, whereas with $k = 1$ we have the most progressive variant of tailoring that does not consider feasibility at all: Since $v_n(o), v_q(o) \in [0, 1]$ we always have $v_n(o) \leq v_q(o) + 1$.

$v_n(o)$	$v_q(o)$	$C(o)$	$rating_C(o)$
0.1	0.2	true	0.1
0.1	0.2	false	0.9
0.1	0.8	true	0.1
0.1	0.8	false	0.9
0.9	0.2	true	0.2
0.9	0.2	false	0.1
0.9	0.8	true	0.9
0.9	0.8	false	0.1

Table 2.2: $rating_C(o)$ with different values for $v_n(o)$ and $v_q(o)$, and feasibility offset $k = 0.2$

Rating Excluded Options

In the case that an option o is excluded from the tailoring configuration, we need not consider its feasibility—it is not any better to exclude an unfeasible option than a feasible one. Instead, we need solely be concerned about its *dispensability* which we define as the one-complement of its necessity:

$$1 - v_n(o) \tag{2.39}$$

A Rating Function for Tailoring Decisions

Pulling it all together, we can now give a rating function for tailoring decisions:

$$rating_C(o) = \begin{cases} \min(v_n(o), v_q(o) + k) & \text{if } C(o) \\ 1 - v_n(o) & \text{if } \neg C(o) \end{cases} \qquad \text{where } 0 \leq k \leq 1 \tag{2.40}$$

Example 2.12 *Table 2.2 lists decision ratings of an option o with different values for $v_n(o)$ and $v_q(o)$, both for the case that o is chosen, and that o is not chosen. As is to be expected, dispensable options ($v_n(o) = 0$) yield a low rating if they are chosen, and a high rating if they are not. Their feasibility does not influence the rating in either case. Options that are both feasible and necessary will be rated high when they are chosen, and low when they are excluded. Options that are necessary but not feasible receive bad ratings in either case.*

If an option is necessary, but not feasible, there is an obvious conflict: It is necessary that the option be chosen, but at the same time it is not feasible to do so. This is reflected by

Option	$rating_{C_1}$	$rating_{C_2}$
o_1	0.4	0.8
o_2	1.0	0.8
o_3	1.0	0.8
$\sum_{o_{1...3}}$	2.4	2.4

Table 2.3: Two example tailoring configurations with different ratings

the fact that neither choosing nor excluding any such option yields a favourable rating. In Table 2.2 from Example 2.12, the rating is slightly better when the option is chosen because of the contribution of the feasibility offset $k = 0.2$.

2.4.4 Rating Tailoring Configurations

Based on the ratings for individual tailoring decisions, we can now calculate a rating for complete tailoring configurations. A straightforward approach is to define the rating of the tailoring configuration simply as the sum of all individual tailoring decisions:

$$rating_{\text{additive}}(C \in \mathbf{C}) = \sum_{o \in O} rating_C(o) \tag{2.41}$$

However, a simple example shows that this rating function does not necessarily lead to desired results:

Example 2.13 *Suppose we have to decide between two tailoring configurations C_1 and C_2 with decisions about a set of three options $\mathbf{O} = \{o_1, o_2, o_3\}$. The decisions are rated as in Table 2.3. If we calculate the ratings of C_1 and C_2 by summing up the two columns, we get $rating_{additive}(C_1) = 2.4$ and $rating_{additive}(C_2) = 2.4$.*

Although the ratings of both configurations are equal, we would intuitively prefer configuration C_2 since every single decision in C_2 has an acceptable rating whereas in C_1 the rating for o_1 is below 0.5 and thus not recommended (see beginning of Section 2.4).

Consequently, we need to modify the rating function such that if two configurations have equal or similar average decision ratings, the configuration is favoured that contains fewer low-rated decisions. This can be achieved by applying bias function

$$f(x) = 1 - (1 - x)^2 \tag{2.42}$$

to every decision rating before calculating the sum of all ratings (Figure 2.6). The idea underlying (2.42) is that for decisions rated below the optimal value of 1, the sum of ratings should not be discounted linearly, but instead by the individual ratings' *square*

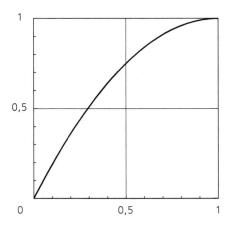

Figure 2.6: Bias function $f(x) = 1 - (1 - x)^2$ for tailoring decision ratings

Option	$f(rating_{C_1})$	$f(rating_{C_2})$
o_1	0.64	0.96
o_2	1.00	0.96
o_3	1.00	0.96
$\sum_{o_{1...3}}$	2.64	2.88

Table 2.4: Biased ratings from Table 2.3

distance from 1. This means that the sum of ratings gets disproportionally lower as ratings of individual tailoring decisions decrease (Table 2.4).

We therefore define the rating function for tailoring configurations as

$$rating(C) = \sum_{o \in O} \left(1 - \left(1 - rating_C(o) \right)^2 \right) \qquad \text{with } C \in \mathbf{C} \qquad (2.43)$$

Figure 2.7 provides an overview of the data flow for calculating the rating for a tailoring configuration: The tailoring context provides measurements for properties of entities. The tailoring configuration contains decisions about tailoring options. Both measurements and decisions form the basis for necessitating and qualifying hypotheses associated with every tailoring option. Every tailoring decision is rated according to the hypotheses of its associated tailoring option. The ratings of all tailoring decisions are then combined to the rating of the tailoring configuration.

2.4.5 Vague Measurements

Up to now, we have based our rating mechanism on simple valuations as defined in Section 2.3.3. However, if we measure properties as ranges (2.13a), we need to substitute every valuation function $g : D \mapsto Q$ that operates on precise measurements by a valuation function $\hat{g} : D \mapsto \hat{Q}$ that operates on measurement ranges:

$$\hat{g}([p, q]) = \begin{cases} [g(p), g(q)] & \text{if } g(p) \leq g(q) \\ [g(q), g(p)] & \text{otherwise} \end{cases} \qquad \text{where } p, q \in D \qquad (2.44a)$$

$$\hat{g}(\bot) = [0, 1] \qquad (2.44b)$$

Note that, as opposed to simple valuations, with (2.44b) we now provide a meaningful definition for the valuation of undefined measurements—the FTI $[0, 1]$ indicates that no information is available.[4]

We then need to valuate the FFs to fuzzy truth intervals instead of fuzzy truth values, using the valuation function $\hat{v}$ from (2.23) for necessity and feasibility:

$$\hat{v}_n(o) = [r_n, s_n] \qquad (2.45a)$$

$$\hat{v}_q(o) = [r_q, s_q] \qquad (2.45b)$$

---

[4] $[0, 1]$ is also the valuation of FIL constant "unknown" (2.23f), see page 22.

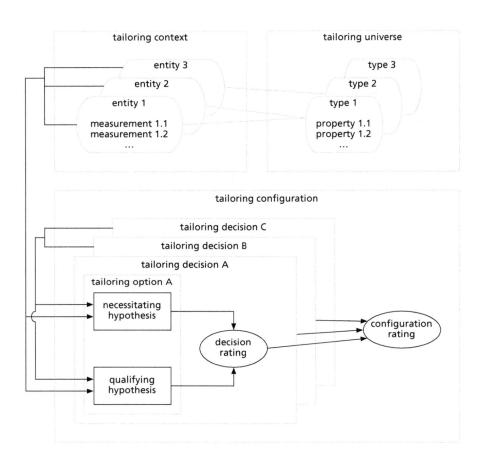

Figure 2.7: Overview of the data flow for calculating the rating for a tailoring configuration

From this, we get an optimistic and a pessimistic rating for tailoring decisions:

$$rating_C^{opt}(o) = \begin{cases} \min(s_n, s_q + k) & \text{if } C(o) \\ 1 - s_n & \text{if } \neg C(o) \end{cases} \tag{2.46a}$$

$$rating_C^{pess}(o) = \begin{cases} \min(r_n, r_q + k) & \text{if } C(o) \\ 1 - r_n & \text{if } \neg C(o) \end{cases} \tag{2.46b}$$

From (2.44) follows that $r_n \leq s_n$ and $r_q \leq s_q$, thus we always have

$$rating_C^{pess}(o) \leq rating_C^{opt}(o) \tag{2.47}$$

To obtain the overall decision rating $rating_C(o)$ we can adopt either only $rating_C^{opt}$ or $rating_C^{pess}$, or might calculate a—possibly weighted—mean of the two values. We actually choose to define $rating_C$ as

$$rating_C(o) = rating_C^{opt}(o) \tag{2.48}$$

because we find it to be of greater value to know the tailoring configuration with the highest potential instead of the one with the lowest risk, that is, the one with the highest *optimistic* rating, considering that in an interactive TSS the user would be able to revise particularly risky tailoring decisions and then have the system re-tailor the remaining options.

Furthermore, we will show in Section 2.5.3 that using the optimistic rating will be essential to our proposition of an efficient algorithm for finding the optimal tailoring configuration.

Rating Partial Tailoring Configurations

In Section 2.4.1 we have shown how tailoring hypotheses can make statements about tailoring decisions, thus introducing dependencies towards the tailoring configuration. By using vague measurements, we can also rate options depending on still undetermined tailoring decisions. We modify valuation function *configured* (2.35) to valuate to intervals and accommodate a third case in which it is yet unknown whether a specific tailoring option is included in the tailoring configuration or not:

$$\widetilde{configured}(o \in \mathbf{O}) = \begin{cases} [1, 1] & \text{if } C(o) \\ [0, 0] & \text{if } \neg C(o) \\ [0, 1] & \text{if } C(o) \text{ is undefined} \end{cases} \quad \text{where } C \in \mathbf{C} \tag{2.49}$$

2.4.6 Tailoring and Resource Planning

In Section 2.4.1 we have shown how FFs can be used to make statements about the tailoring context with the aid of FPV. In Example 2.11 we have based a valuator on the measurement for a specific entity. Yet when defining hypotheses for tailoring options, we will usually not be able to refer to specific entities as these will be different in every tailoring context, and because no concrete entities might have been defined at the time the hypotheses are written. Instead, we will have to generalise our statements to refer to arbitrary representatives of entity *types*.

Example 2.14 *The valuator from Example 2.11, generalised to types, is*

$$v_{big\text{-}team} = ramp_{5,10}(measurement(e, \text{team-size})) \quad with\ e \in \mathbf{E}, e\text{-}type(e) = \text{project}$$

The valuator in Example 2.14 is unambiguous as long as only one entity of type "project" is known. However, in tailoring context **K** there can be multiple entities of the same entity type. Since the entity type determines the properties of the associated entities, some properties may be defined more than once. In Example 2.6 on page 17, entities "peter" and "laura" are both of type "staff" and thus both define an "experience" property.

Put formally, we have to deal with the situation when we have n entities $e_1, \ldots, e_n$ of the same type $t = e\text{-}type(e_{1\ldots n})$ and for the same property p, and a general FPV referring to measurements of that property p of all entities of type t.

Example 2.15 *Suppose that option o represents a project management activity. The rating of the option depends on the "experience" property from Example 2.6. However, we have no means of telling whether Peter or Laura will eventually be responsible for managing the project. So which measurement of the two experience properties should be used to evaluate the rating?*

We can use fuzzy interval logic and range measurements to escape the dilemma of multiple measurements. Consider a function $\hat{g}$ that calculates FPV valuations from range measurements of property p. If we have n entities $\mathbf{E} = \{e_1, \ldots, e_n\}$ with that property, we can apply $\hat{g}$ to all range measurements $\hat{m}_1, \ldots, \hat{m}_n \in \{[u, v] \mid u, v \in \mathbf{D}, u \leq v\}$ of p and obtain a list of FTIs $\hat{g}(\hat{m}_1), \ldots, \hat{g}(\hat{m}_n) \in \hat{\mathbf{Q}}$. We now need to reduce this set of interval valuations to a single interval that marks the optimistic and pessimistic valuations of the *set* of entities $\mathbf{E}$.

There is not a unique way of calculating a valuation for a set of range measurements that is appropriate under all circumstances. Consider the following examples:

Example 2.16 (**"weakest link in the chain" approach**) *Choosing* Java *as programming language for a new software development project requires all developers to have sufficient experience with the language. The team's capability is considered equal to the capability of the least experienced developer. Hence, the optimistic capability estimate amounts to the lowest optimistic estimate of any one team member, and the pessimistic capability estimate amounts to the lowest pessimistic estimate of any one team member.*

Example 2.17 (**"one for all" approach**) *One team member has to take the role of the requirements engineer, who needs to be capable of a certain level of structured and formal writing. Thus the feasibility of tailoring option "appoint requirements engineer" is determined by the capability valuation of the appointed team member. Since we can make no assumptions about role allocations, we have to assume that any team member could be appointed. Thus the optimistic valuation is equal to the highest optimistic capability valuation of any one team member, and the pessimistic valuation is equal to the lowest pessimistic capability valuation of any one team member.*

There is also a middle course between examples 2.16 and 2.17 that we put forward in the next example:

Example 2.18 (**"committee" approach**) *A certain quality assurance method requires two team members to take responsibility. Its feasibility depends on the truth of the statement "appointed two team-members need to have been in the company for approximately three years at minimum." Again, it is unknown which two team members will actually be appointed, so in the face of that uncertainty we have to broaden the statement to "any selection of two team-members needs to have been in the company for approximately three years at minimum."*

The worst case is that among the two appointed team-members, one will have been in the company for the shortest time of all team-members. The best case is that the two team-members who have been in the company for the longest time will be appointed.

Hence, applying the weakest link *principle among two randomly picked appointed team-members, the pessimistic valuation of the above statement is once more equal to the lowest pessimistic valuation of any single team member (i. e., the team member who has been in the company for the shortest time), whereas the optimistic valuation is equal to the second-highest optimistic valuation of any single team member (i. e., the team member who has been in the company for the second-longest time).*

While Example 2.18 adopts the *weakest link* approach from Example 2.16, we limit it to a smaller subset of the entities in question. From that point of view, Example 2.17 represents the *weakest link* approach applied to subsets of size one.

We can draw a parallel between examples 2.16 through 2.18. For n entities with individual valuations $\hat{g}(\hat{m}_1), \ldots, \hat{g}(\hat{m}_n)$ for the FPV in question, the overall pessimistic valuation is always the lowest pessimistic valuation of any single entity, whereas the optimistic valuation is the c-highest optimistic valuation of any single entity:

$$
\hat{g}(\hat{m}_1, \ldots, \hat{m}_n) = \begin{cases} [0, 0] & \text{if } n \geq 0, c > n \\ [0, 1] & \text{if } c = n = 0 \\ \left[\min_{1 \leq i \leq n} r_i, \max^c_{1 \leq i \leq n} s_i\right] & \text{otherwise} \end{cases} \tag{2.50}
$$

with

$$
[r_i, s_i] = \hat{g}(\hat{m}_i) \tag{2.51a}
$$

$$
\max^c\{x_1, \ldots, x_n\} = x_{n+1-c} \qquad \text{where } \forall i, j \in \{1, \ldots, n\} : i < j \Rightarrow x_i \leq x_j \tag{2.51b}
$$

For the *weakest link* approach, we have $c = n$, for the *one for all* approach, we have $c = 1$ and for the *committee* approach we have $c > 1$.

Whenever we prescribe a fixed value for c, we have to deal with the case that $c > n$, i.e., there are fewer entities than are required. In that case, $\overset{c}{\max}$ is not defined; and the FPV should valuate to $[0, 0]$ for lack of sufficient entities.

In the case that $c = n = 0$, no definitive statement can be made about the FPV, therefore it must valuate to $[0, 1]$.

Fuzzy Propositional Variables about Entity Types

Based on (2.50), we now define generic FPVs on tailoring contexts along with their associated valuators. Suppose we have a valuation function for measurements $g : D \mapsto \hat{Q}$, an entity type $t \in T$ and a property $p \in P$. Then the set of all range measurements for type t and property p is

$$
\hat{M}_{t,p} = \bigcup_{\{e \in E | e\text{-}type(e) = t\}} \widetilde{measurement}(e, p) \tag{2.52}
$$

where

$$
\hat{M}_{t,p} \subseteq \{[u, v] \mid u, v \in D, u \leq v\} \cup \{\bot\} \tag{2.53}
$$

In parallel with the *weakest link* aproach from Example 2.16 we define a generic FPV as

$$
proposition(t, p, g) \tag{2.54}
$$

and we define the associated valuation function as follows:

$$\hat{v}(proposition(t, p, \hat{g})) = \begin{cases} [0, 1] & \text{if } \hat{M}_{t,p} = \emptyset \\ \left[\min_{m \in \hat{M}_{t,p}} lo\left(\hat{g}(\hat{m})\right), \min_{m \in \hat{M}_{t,p}} hi\left(\hat{g}(\hat{m})\right) \right] & \text{otherwise} \end{cases}$$

$$(2.55)$$

where $\hat{g}$ is derived from g in (2.54) as defined in (2.44).

Example 2.19 *We can define FPV big-team from Example 2.11 in terms of generic proposition (2.54):*

$$big\text{-}team = proposition(\text{project, team-size}, ramp_{5,10}) \qquad (2.56)$$

It can be valuated with valuator (2.55).

To introduce a cardinality constraint $c \in \mathbb{N}$ as required for examples 2.17 and 2.18, we write a FPV in the generic form

$$proposition(t, p, g, c) \qquad (2.57)$$

and valuate it as

$$\hat{v}(proposition(t, p, \hat{g}, c)) = \begin{cases} [0, 0] & \text{if } c > |\hat{M}_{t,p}| \\ \left[\min_{\hat{m} \in \hat{M}_{t,p}} lo\left(\hat{g}(\hat{m})\right), \overset{c}{\max_{\hat{m} \in \hat{M}_{t,p}}} hi\left(\hat{g}(\hat{m})\right) \right] & \text{otherwise} \end{cases} \qquad (2.58)$$

Example 2.20 *The statement "any selection of two team-members needs to have been in the company for approximately three years at minimum" from Example 2.18 can be expressed as generic FPV*

$$experienced\text{-}3\text{-}years = proposition(\text{staff, experience}, ramp_{30,42}, 2) \qquad (2.59)$$

where property "experience" is measured in man-months. experienced-3-years can be valuated with valuator (2.58)

2.5 The Optimisation Algorithm

We will now provide an algorithm that delivers an optimal tailoring configuration with respect to the definitions given in Sections 2.2 to 2.4. To this purpose, we will first abstract the tailoring problem as far as possible in order to gain an unobstructed perspective of the inner workings of the algorithm.

2.5.1 The Optimisation Problem

We can summarise the optimisation problem as follows:

Tailoring Problem Definition 1 *Given*

- *a set of* binary options **O**,

- *the set of* tailoring configurations $\mathbf{C} = \mathbb{B}^{\mathbf{O}}$, *i. e., the family of mappings from options in* **O** *to boolean values* $\mathbb{B}$, *and*

- *a* rating function

$$rating : \mathbf{C} \mapsto \mathbb{R},$$

find the tailoring configuration $C^* \in \mathbf{C}$ *such that*

$$rating(C^*) = \max_{C \in \mathbf{C}} rating(C). \tag{2.60}$$

In other words, we have a *search problem*: we have a search space **C**, and we have a characterisation of the item we are looking for. For example, $\mathbf{O} = \{a, b\}$ would require us to search the best tailoring configuration out of

$$\mathbf{C} = \mathbf{C}_{\{a,b\}} = \{\{a \rightarrow \text{false}, b \rightarrow \text{false}\}, \{a \rightarrow \text{false}, b \rightarrow \text{true}\},$$
$$\{a \rightarrow \text{true}, b \rightarrow \text{false}\}, \{a \rightarrow \text{true}, b \rightarrow \text{true}\}\}.$$

For $|\mathbf{O}| = n$ options there are $|\mathbf{C}| = 2^n$ possible mappings. Therefore, a *brute force* approach of evaluating all mappings and keeping the best mapping would entail exponential time complexity. We will therefore propose a more sophisticated strategy to find a solution. We will outline our approach in two parts: First, we will provide a suitable structure for the search space **C**, and then we will develop a search algorithm based on a variant of the A* shortest-path search algorithm.

2.5.2 The Search Space

Up to now, we have an unstructured search space **C**. In order to be better than the brute-force method mentioned above, we will have to examine only a part of the search space while making sure that we have not overlooked the tailoring configuration $C^* \in \mathbf{C}$ with the best rating.

To this end, we will develop an incremental approach, working our way towards a complete tailoring configuration, decision by decision. We will therefore need to consider all subsets of **O** of size i:

$$\mathbf{P}_i^{\mathbf{O}} = \{\mathbf{O}' \mid \mathbf{O}' \subseteq \mathbf{O} \wedge |\mathbf{O}'| = i\} \tag{2.61}$$

Thus, for $|O| = n$ we have

$$P_0^O = \{\emptyset\}$$
$$P_1^O = \{\{o_1\}, \{o_2\}, \dots, \{o_n\}\}$$
$$P_2^O = \{\{o_i, o_j\} \mid 1 \le i, j \le n \land i \ne j\}$$
$$\vdots$$
$$P_n^O = \{O\} \tag{2.62}$$

and

$$\bigcup_{0 \le i \le n} P_i^O = \mathcal{P}(O) \tag{2.63}$$

For any option subset $O' \in \mathcal{P}(O)$ we have $2^{|O'|}$ partial configurations $C_{O'} \in \mathbf{C}_{O'} = \mathbb{B}^{O'}$. Taking all of these partial configurations together, we define the enhanced configuration space

$$\mathbf{C}^* = \bigcup_{O' \in \mathcal{P}(O)} \mathbf{C}_{O'} \tag{2.64}$$

If, for example, we have options $O = \{a, b\}$, then the enhanced configuration space is

$$\begin{aligned}
\mathbf{C}^* = \{&\emptyset, \{a \to \text{false}\}, \{a \to \text{true}\}, \{b \to \text{false}\}, \{b \to \text{true}\}, \\
&\{a \to \text{false}, b \to \text{false}\}, \{a \to \text{false}, b \to \text{true}\}, \\
&\{a \to \text{true}, b \to \text{false}\}, \{a \to \text{true}, b \to \text{true}\}\}.
\end{aligned} \tag{2.65}$$

We now add structure to the configuration space by introducing edges $\mathbf{E}_{\mathbf{C}^*} \subseteq \mathbf{C}^* \times \mathbf{C}^*$ where each edge $(C_1, C_2) \in \mathbf{E}_{\mathbf{C}^*}$ is required to connect two configurations C_1 and C_2 where the C_2 differs from the C_1 only in that it includes one additional option and maps to the same values for all other options as in C_1. Figure 2.8 illustrates the search graph for the above example configuration space (2.65).

Formally, the set of edges $\mathbf{E}_{\mathbf{C}^*}$ is defined as

$$\mathbf{E}_{\mathbf{C}^*} = \{(C_1, C_2) \mid (\exists O' \subset O, o \in O)(C_1 \in \mathbf{C}_{O'} \land C_2 \in \mathbf{C}_{O' \cup \{o\}} \land C_1 \subset C_2)\} \tag{2.66}$$

We can consider our search space complete as long as there are paths from the empty configuration $C_\emptyset$ to all complete configurations in $\mathbf{C}$. For the example in Figure 2.8, we can thin out the graph in two ways without losing any of the four complete configurations. Figure 2.9 shows the resulting search trees.

Formally, this requires us to prescribe an order in which the options in O are to be included in the growing configuration space. This can be achieved by defining a function

$$next : \mathbf{C}^* \mapsto O \tag{2.67}$$

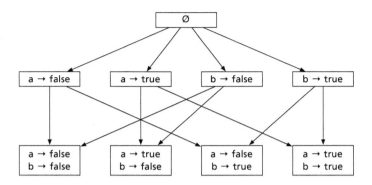

Figure 2.8: The search graph for the configuration space from (2.65).

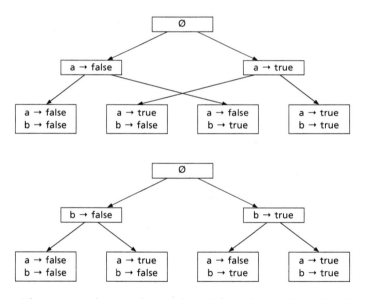

Figure 2.9: The two complete search trees derived from the search graph in Figure 2.8.

where for any $C \in \mathbf{C}_{O'}$ we have

$$next(C) \in \mathbf{O}\backslash\mathbf{O}', \tag{2.68}$$

i.e., function *next* proposes only options that are not already considered in the given configuration space. Because of this, function *next* is not completely defined because there are no next options for complete configurations $C \in \mathbf{C}$.

Our reduced set of edges is thus

$$\mathbf{E}_{\mathbf{C}^*}^- = \{(C_1, C_2) \mid (\exists \mathbf{O}' \subset \mathbf{O})(C_1 \in \mathbf{C}_{O'} \wedge C_2 \in \mathbf{C}_{O' \cup \{next(o)\}} \wedge C_1 \subset C_2)\} \tag{2.69}$$

We can now restate the optimisation problem from Section 2.5.1:

Tailoring Problem Definition 2 *In the search tree*

$$\mathbf{S} = (\mathbf{C}^*, \mathbf{E}_{\mathbf{C}^*}^-) \tag{2.70}$$

find the path from root $C_\emptyset$ to configuration $C^* \in \mathbf{C}$ such that

$$rating(C^*) = \max_{C \in \mathbf{C}} rating(C). \tag{2.71}$$

2.5.3 The Search Algorithm

The search tree as particularised in Section 2.5.2 allows us to take an incremental approach to finding an optimal configuration. By examining partial configurations, we expect to make predictions about which areas of the search space should be considered first, and, more importantly, which areas can be safely excluded from search without putting the optimal solution out of reach. Thus, we need to extend the domain of our rating function *rating* to also include partial configurations:

$$rating : \mathbf{C}^* \mapsto \mathbb{R} \tag{2.72}$$

We will refer to ratings of partial configurations as *partial ratings*.

The choice of search strategy will depend on the properties of the partial ratings that our rating function returns. We will first examine the properties required for a successful best-first search, and will then go on to show how an adapted A* algorithm can succeed when the requirements on partial ratings need to be further relaxed.

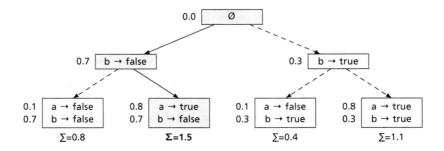

Figure 2.10: Best-first search with a distributive rating function. The numbers indicate the ratings for the individual decisions about options a and b.

Best-First Search

We say to have a *distributive* rating function if the individual ratings of any two partial configurations add up to the rating of their union:

$$(\forall C_1, C_2 \subseteq \mathbf{C})\,(C_1 \cap C_2 = \emptyset \implies rating(C_1 \cup C_2) = rating(C_1) + rating(C_2)) \quad (2.73)$$

Having a distributive rating function means that decisions about single options have no impact on the ratings of other decisions; we can therefore optimise each decision independently. This is because from (2.73) follows that for any configuration $C = \{c_1, c_2, \ldots, c_n\}$ we have

$$rating(C) = rating(\{c_1\}) + rating(\{c_2\}) + \cdots + rating(\{c_n\}). \quad (2.74)$$

Since we have only binary decisions, optimising a decision means picking the choice that gets us the better rating of the two.

Our search strategy in this case is a straightforward best-first search: Descending through the search tree starting from the root configuration, always follow the branch that leads to the configuration with the best rating. The key idea is that by picking the best choice available locally at any time, we are guaranteed to get to the best global solution. See Figure 2.10 for an example.

However, as soon as ratings of individual decisions are not independent but also depend on other decisions, we no longer have a distributive rating function, and the above approach will in most cases not lead to the best configuration, as illustrated in Figure 2.11. All the worse, non-distributivity entails that partial ratings cannot be expected to give any indication whatsoever about which branch is to be preferred, so even if we introduce backtracking best-first search it is no better than a random walk through the tree.

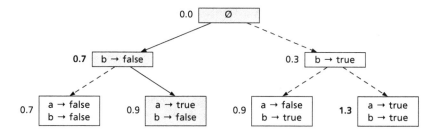

Figure 2.11: If ratings are not distributive, best-first search does not guarantee to lead straight to the best result.

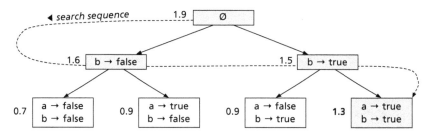

Figure 2.12: Configuration tree with optimistic ratings. The T* algorithm starts at the root and then continues its search along the dotted line. The first leaf node that it picks is guaranteed to have the best rating of all leaf nodes.

Optimistic Search

We have developed a search algorithm that does not depend on a distributive rating function as required in the preceding section. It is derived from the A* algorithm [Wika], and in the following we will refer to it as T*. Figure 2.12 illustrates an example of its workings. The ratings of the partial configurations have a specific property: The rating of any partial configuration is never worse than the best partial configuration reachable from there by descending the tree, i.e., by adding further decisions to that partial configuration. We talk of an *optimistic* rating function if it has precisely this property. The intuition behind this is that every rating represents an optimistic estimate of the best rating that can be achieved with the decisions yet to be made. Formally, an optimistic rating function must satisfy the following condition:

$$(\forall C_1, C_2 \in \mathbf{C}^*)\,(C_1 \subseteq C_2 \implies rating(C_1) \geq rating(C_2)) \qquad (2.75)$$

This is also the minimum requirement for the T* search algorithm to succeed: When searching the tree, T* maintains a set of *open nodes* N. Initially, N contains only the root node. In each iteration of the search, T* picks from N the node n^* with the highest rating, i.e., $rating(n^*) = \max_{n \in N} rating(n)$. If T* has picked a leaf node, the search ends and we are guaranteed to have found the best overall leaf node in the tree. Otherwise, T* removes n from N and instead adds all child nodes of n to N. The open nodes set N is usually implemented as a priority queue, as this is efficient both for adding new nodes and for retrieving and removing the currently best node.

In our context, nodes in the tree are configurations $C \in C^*$, and child nodes are determined by virtue of the reduced edge set $E_{C^*}^-$ as defined in (2.69).

Given that the rating function is optimistic, the T* algorithm

1. always finds the best leaf node, and

2. it does so by examining fewer nodes than any other algorithm using the same rating function.

The intuition behind these two claims is as follows: When T* terminates its search, the leaf node n^* it has picked from N has a better rating than any other node in N. But since the ratings are optimistic, and are all below the rating of n^*, we know for sure that further search will not eventually reveal a leaf node with a better rating than n^*. This justifies claim 1.

For claim 2, it is necessary to understand that when T* terminates it will have examined all nodes in the tree that have a higher rating than the resulting node, and no nodes with a lower rating. Consider now an algorithm that examines the nodes in the tree in some other order and eventually picks out a leaf node n after having examined fewer nodes than the above algorithm. From this follows that there is at least one unexamined node n' in the tree with a better rating than n. Since the rating function is optimistic, it is possible that some other leaf node below n' has a better rating than n. Hence, T* is guaranteed to examine the minimum number of nodes necessary in order to make certain that the optimal leaf node is found.

The remaining factor for improving the search algorithm is thus the rating function. The closer the optimistic ratings get to the actual best rating achievable from any partial configuration, the better the search algorithm is at discerning the most promising branches in the search tree. Nevertheless, the calculation of the optimistic rating function for any node should not extensively examine the sub-tree rooted at that node, since that would introduce a sub-search in its own right and would thus undermine the efficiency of the search algorithm. So the ideal rating function, which for each node predicts the exact rating of its best descendant leaf node, is in most cases unrealistic.

An Optimistic Rating Function

In Section 2.4.5, we have already given an optimistic rating function based on hypotheses expressed as fuzzy formulæ. Using vague measurements, we have shown that we could valueate the FFs to fuzzy truth intervals. From these, we could derive an optimistic rating (2.46a). This rating is also optimistic with regard to FFs containing FPVs valuated with function $\widehat{configured}(o)$ (2.49).

Function $\widehat{configured}$ refers to the tailoring configuration to be calculated by the optimisation algorithm, and, unlike other valuation functions, is subject to change as the optimisation algorithm proceeds. For options that have not yet been decided about, $\widehat{configured}$ returns the maximally vague FTI $[0, 1]$. Therefore, augmenting a partial tailoring configuration by a decision about a yet undecided option can never improve the optimistic rating nor can it lower the pessimistic rating; hence, the optimality criterion cannot be violated.

Parallels between T* and A*

As noted above T* is based on the A* algorithm. Like T*, A* uses optimistic estimates in order to find the best result with minimal computational effort. There are several differences, though:

- A* is designed to search the shortest path between two distinct nodes in a graph. It does so by minimising an optimistic cost function. T*, however, operates on trees; it maximises a rating function, and its goal is not a predetermined node but whichever leaf node that is reached first.

- A* in many settings needs to maintain a list of *closed nodes* to prevent nodes from being visited more than once via different paths. Since T* operates only on trees, it can do without this additional bookkeeping.

- A* relies on a graph with weighted edges. It calculates the estimated cost of the best path through node n by estimating the cost of the remaining, yet unknown path from n to the goal node, and adding to that the known cost of the path already covered up to node n. T* only requires an optimistic estimate of the rating of the best leaf node reachable from n and can do without the additional bookkeeping required for A* to keep track of costs of partial paths.

2.5.4 Improving the Search Algorithm

T* terminates when the first leaf node has been reached in the tree. In the worst case, the priority queue will grow until it contains all nodes one level above the leaf nodes, before a leaf node is finally selected. Hence, with n options, the queue may have to accommodate as many as 2^{n-1} elements, and up to 2^n will have to be searched. Figure 2.12 is an example of a worst case for $n = 2$ options.

Even if this worst case is unlikely to happen in practice with more but a few tailoring options, we need to take precautions against T* using excessive time and memory resources in order to come to terms with the potentially exponential growth of resources required with regard to the problem size. We have combined several approaches to this end, each of which we will outline in the remainder of this section: First, we will show how the tailoring problem can be broken down to smaller subproblems. Then, we will discuss how the order in which tailoring options are considered can contribute to finding the optimal solution more quickly. Finally, as a last resort, we will propose a heuristic to prune unpromising branches from the search tree without significantly—or even necessarily—compromising the search result.

Partitioning the Search Space

As stated above, we cannot decide about each tailoring option independently because there may be dependencies between tailoring options. For this reason the rating function is not distributive and we cannot use the (linear) best-first search algorithm. Instead, we apply the T* algorithm whose worst-case time and memory complexity is exponential.

In most cases, however, not every option depends, directly or indirectly, on every other option. Rather, we can expect the tailoring universe to contain several clusters of interdependent options. If we partition the search space according to these clusters and calculate the tailoring configuration partition by partition, we can lower the overall complexity significantly (Figure 2.13): Instead of one search problem of size n, we have k search problems of sizes $m_1, \ldots, m_k$ with $m_1 + \ldots + m_k = n$ and a maximum of $\sum_{1 \leq i \leq k} 2^{m_i}$ nodes to be searched across all trees, which is significantly less than 2^n for a single search tree, even if the largest partition is of size $n - 1$.

Improving the Node Sequence

In Section 2.5.2 we have introduced function *next*, which determines the order in which tailoring options are evaluated during the search for the optimal tailoring configuration. Given a set of tailoring options O with size $|O| = n$, we can generate $n!$ permutations of O. Function *next* produces one of these orderings, and thus determines which level of the search tree is associated with which tailoring option. Figure 2.9 provides an example for the two permutations that are possible with $O = \{a, b\}$, namely (a, b) and (b, a).

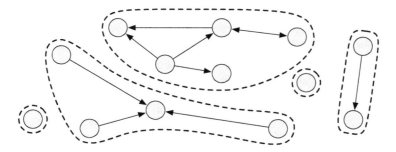

Figure 2.13: Clusters of interdependent tailoring options. The circles represent options, the arrows indicate dependencies.

The incremental nature of the T* algorithm does not require us to completely determine the node sequence in advance. Instead, function *next* will ever only need to determine the immediately next option to be evaluated.

The simplest implementation of function *next* is to have it pick a random option from the set of yet unprocessed options:

$$next_{\text{random}}(C \in \mathbf{C}_{O'}) = \operatorname*{random}_{o \in O \setminus O'} o \tag{2.76}$$

However, this approach does not respect the fact that the order of options delivered by function *next* influences the total number of nodes that need to be examined during a T* search. As we will account for further below, a well chosen node sequence can greatly reduce the total number of nodes searched. Yet we need to bear in mind that the heuristics for picking the most promising option must be computationally simple in order not to undermine the time and space complexity of the T* algorithm. Function *next* must not engage in a complex search by itself.

Before we propose a heuristic for function *next*, let us recall the part it plays in the T*algorithm. In every iteration, T* removes the node with the highest rating from its set of open nodes. It then generates all successors of that node and places them in the set of open nodes. With nodes representing partial tailoring configurations $C \in \mathbf{C}_{O'}$, there are always two successors: Function *next* will pick a yet undecided tailoring option o, and the successors will be partial tailoring configurations extended by a decision about o:

$$C_1^o = C \cup \{o \rightarrow \text{true}\} \qquad \text{and} \qquad C_2^o = C \cup \{o \rightarrow \text{false}\} \tag{2.77}$$

This relation between the original partial configuration C and the derived configurations C_1^o and C_2^o is reflected in the set of edges in the search tree, $\mathbf{E}_{\mathbf{C}^*}^-$ (2.69).

The heuristic we apply for function *next* is similar to the approach adapted in many expert systems [FN86]: At every iteration of the T* algorithm, we extend the tailoring configuration by the option o^* that brings the greatest increase in information.

How can we determine the increase in information? Recall that T* uses an optimistic rating function. The rating of any partial tailoring configuration $C \in \mathbf{C}^*$ thus reflects the upper bound for the ratings of all complete tailoring configurations incorporating C. Now consider a partial tailoring configuration C that has just been removed from the set of open nodes. For every tailoring option o not yet considered, we can generate two succeeding tailoring configurations C_1^o and C_2^o as defined above. We can then calculate the optimistic ratings for both C_1^o and C_2^o. The greater one of these two ratings will give us a new upper bound of the rating that can be achieved for tailoring configurations incorporating C:

$$rating^o(C) = \max_{i \in \{1,2\}} rating(C_i^o) \tag{2.78}$$

Since we have an optimistic rating function, we have

$$rating^o(C) \leq rating(C) \qquad \text{with } C \in \mathbf{C}_{O'}, o \in O \backslash O' \tag{2.79}$$

Thus, by looking ahead by one decision, we can improve our optimistic prediction of the best final rating achievable for partial configuration C. In order to improve this prediction as much as possible, we need to extend our current partial tailoring configuration C by option o^* that minimises $rating^o$:

$$next(C \in \mathbf{C}_{O'}) = o^* \qquad \text{where } rating^{o^*}(C) = \min_{o \in O \backslash O'} rating^o \tag{2.80}$$

Thus, we always choose the tailoring option that restricts our (optimistic) expectations of the final rating as far as possible. Thereby we prevent useless searches of subtrees that promise better final ratings than we could knowingly obtain.

Beam Search: Pruning the Search Tree

Both partitioning the search space and improving the node sequence are methods that aim at reducing the complexity of the search problem. Yet they cannot fully rule out the possibility that the search space grows beyond practical time and space boundaries. Therefore, as a last resort, we need to provide the means of pruning branches from the search tree.

To achieve this, we limit the size of the set of open nodes N maintained by the T* algorithm (see page 43). When, after adding a new node to N, the size limit has been exceeded, the node with the worst rating—hence, the lease promising node—is subsequently removed from N. This approach to pruning the search space is known as *beam search* [FN86].

Since at the end of the search N will usually not be empty, there is a good chance that the removed nodes would not even have been considered, provided that the size limit for N had been high enough. In this case, the optimal solution is found even though nodes have been removed from N.

In order to tell whether the solution found by T^* is optimal although nodes have been removed from N, T^* needs to keep track of the highest rating of all nodes removed from N due to the size limit.

As long as this rating is lower than the rating of the leaf node resulting from the search, we are still guaranteed to have found the optimal solution: Since the principle of optimistic ratings entails that no node in N will ever be replaced by child nodes with better ratings, and since all nodes in N are processed in decreasing order of their ratings, we can tell that none of the removed nodes would ever have been considered if they would not have been prematurely removed from N.

If, however, the best removed node was rated higher than the resulting node delivered by T^*, we can at least specify the maximum distance to the potentially missed optional solution. Only if this distance is significant, a second T^* search should be performed with more space allocated for N.

An interesting property of beam search is that if we limit the size of N to only one entry, T^* will behave like the best-first search algorithm.

3 Improving the Practicality of the Tailoring Framework

In Chapter 2 we have developed a tailoring framework, including a formal model of tailoring and an optimisation algorithm for tailoring configurations. The concepts we have introduced are all but trivial. Therefore, to apply the framework successfully in a practical TSS, there must be appropriate interfaces for two basic types of users:

1. the *process modeler* who uses the TSS to set up and edit tailoring guidelines, and

2. the *process tailorer* who uses the TSS to do the actual tailoring.

Among the process modeler's tasks are defining a tailoring universe and identifying tailoring options, but the most complex task the process modeler is faced with is expressing the conditions determining acceptable tailoring configurations, especially when complex dependencies between tailoring options have to be modelled: According to our tailoring framework, tailoring conditions have to be expressed individually per option in the form of necessitating and qualifying hypotheses. In the case of mutual dependencies between different tailoring options, this means that hypotheses have to be written separately for every tailoring option involved in the dependency. In Section 3.1 we show how a TSS can help the process modeler overcome this difficulty by allowing him to express such conditions as a single, more general tailoring rule which the TSS then automatically transforms to equivalent tailoring hypotheses.

The process tailorer provides data about the tailoring context by supplying estimates and measurements about entities in the tailoring context. He can also manually choose or exclude some tailoring options before he invokes the tailoring algorithm of the TSS, which then completes the tailoring configuration by providing recommendations for all remaining tailoring decisions. The TSS also provides ratings for all tailoring decisions, regardless of whether a decision has been made by the process tailorer or recommended by the tailoring algorithm. The process tailorer then reviews the rated tailoring configuration and either accepts it, or changes some decisions and then starts another tailoring iteration, allowing the tailoring algorithm to reconsider all tailoring decisions that have not yet been changed or acknowledged manually by the process tailorer. This interaction requires that rated tailoring configurations are presented as clearly and intuitively as possible to the process tailorer, and that the process tailorer is able to understand the reasons for tailoring decisions recommended by the TSS. Section 3.2 shows how the TSS can provide more insight into its tailoring recommendations by displaying ratings not only for tailoring decisions, but also for measurements and estimates from the tailoring context provided by the process tailorer. With the help of structured justifications for both kinds of ratings, the TSS also reveals to the process tailorer what facts and dependencies have contributed to each rating.

3.1 Defining Tailoring Rules

The goal of software-assisted tailoring is to find an optimal tailoring configuration. The optimal tailoring configuration is, by our definition, the tailoring configuration with the highest rating among all possible tailoring configurations. As illustrated in Figure 2.7 on page 32, the rating of a tailoring configuration is calculated from the ratings of its constituting tailoring decisions. Each tailoring decision is about a tailoring option and is rated based on the fuzzy valuations of tailoring hypotheses that have been associated with that tailoring option.

We have introduced two types of tailoring hypotheses: A *necessitating* hypothesis determines under which circumstances a tailoring option is necessary, and a *qualifying* hypothesis determines under which circumstances a tailoring option is feasible (see Section 2.4.2 on page 26).

A TSS based on our tailoring framework ultimately expects the process modeler's knowledge about tailoring to be expressed in terms of these tailoring hypotheses. In order to make automated tailoring assistance as effective as possible, the engineer must be able to translate his knowledge into these formal hypotheses as directly and as intuitively as possible. We will illustrate in Section 3.1.2 that there are cases in which it is much easier to express tailoring rules not in terms of hypotheses about individual tailoring options, but in the form of more general rules. On the other hand, in the interest of not introducing additional complexity, we do not want to extend the formal underpinnings of core tailoring model. Therefore, in sections 3.1.3 and 3.1.4, we will develop a mechanism to transform general tailoring rules to an equivalent set of hypotheses for individual tailoring options. To provide a sound theoretic foundation for this transformation, we require a better understanding of the semantics of tailoring hypotheses. We will show in Section 3.1.1 that tailoring hypotheses can be expressed in terms of deontic logic, which adds a modal *obligation* operator to standard propositional logic. Deontic logic will supply the formal foundation for the rule transformations in sections 3.1.3 and 3.1.4.

Definitions

Before we proceed, we give some definitions that will be helpful in the following sections. Let $A \subset F$ denote the set of all *atomic fuzzy formulae* in F. We restrict A to contain only

FPVs of types *configured* (see (2.49) on page 33) and *proposition* (see (2.54) and (2.57) on page 36), and the three standard FIL constants (see (2.23) on page 22):

$$\mathbf{A} = \{configured(o) \mid o \in \mathbf{O}\} \cup \{proposition(\ldots)\} \cup \{\text{true}, \text{false}, \text{unknown}\} \qquad (3.1)$$

We will use symbols ϕ, ψ, ω to denote arbitrary FFs in $\mathbf{F}$, and symbols α, β to denote atomic FFs in $\mathbf{A}$.

We will also make use of iterative variants of logic operators $\wedge$ and $\vee$:

$$\bigwedge_{\phi \in \{\phi_1, \ldots, \phi_n\}} \phi =_{\text{def}} \phi_1 \wedge \ldots \wedge \phi_n \qquad (3.2a)$$

$$\bigvee_{\phi \in \{\phi_1, \ldots, \phi_n\}} \phi =_{\text{def}} \phi_1 \vee \ldots \vee \phi_n \qquad (3.2b)$$

Iterations over an empty set yield the operators' respective identity elements:

$$\bigwedge_{\phi \in \emptyset} \phi =_{\text{def}} \text{true} \qquad (3.2c)$$

$$\bigvee_{\phi \in \emptyset} \phi =_{\text{def}} \text{false} \qquad (3.2d)$$

3.1.1 Tailoring Rules in Deontic Logic

To introduce the concepts underlying deontic logic, we will take a new perspective on tailoring hypotheses. A tailoring hypothesis is defined by a FF whose atoms are FPVs in $\mathbf{A}$—either propositions about the tailoring context such as *proposition*(project, team-size, $ramp_{5,10}$) (Example 2.19 on page 37), or propositions about the tailoring configuration such as *configured*(use-cases), or FIL constants such as 'unknown.'

A necessitating hypothesis $hyp_{\text{n}}(o) = \phi$ for a tailoring option o has the interpretation

> It is obligatory that if ϕ holds, then option o is configured. $\qquad (3.3a)$

A qualifying hypothesis $hyp_{\text{q}}(o) = \psi$ about option o is the converse of (3.3a) and has the interpretation

> It is obligatory that if option o is configured, then ϕ holds. $\qquad (3.3b)$

Expressing Rules in Deontic Logic

If we represent the phrase "it is obligatory that ..." with an operator O, we can rewrite statement (3.3a) as the logic term

$$O(\phi \rightarrow configured(o)) \tag{3.4a}$$

and statement (3.3b) as

$$O(configured(o) \rightarrow \psi) \tag{3.4b}$$

and thus get a new, alternative notation for hypotheses $hyp_n(o) = \phi$ and $hyp_q(o) = \psi$.

With O we have introduced a modal operator that pertains to a variant of modal logic known as *deontic logic* [Garo5]. Deontic logic comes to play when rules are expressed, that is, when statements are put forward that *ought* to be the case. Intuitively, operator O introduces an *obligation*, as is the case with the necessitating and qualifying tailoring hypotheses defined in (2.36): For tailoring option $o \in O$ and hypotheses $hyp_n(o), hyp_q(o) \in F$ we have

$$O(hyp_n(o) \rightarrow configured(o)) \land O(configured(o) \rightarrow hyp_q(o)) \tag{3.5}$$

We call O ϕ a *rule*, and refer to the subordinate term ϕ as the *condition* of the rule.

Normalised Rules

With the deontic operator we can express rules of arbitrary complexity. Within the context of tailoring rules, we are particularly interested in two special cases of rules which we refer to as *normalised rules*. We define normalised rules to be rules of the form

$$O(\phi \rightarrow \alpha) \qquad \text{with } \alpha \in A, \phi \in F \tag{3.6a}$$

and

$$O(\beta \rightarrow \psi) \qquad \text{with } \beta \in A, \psi \in F \tag{3.6b}$$

We call (3.6a) a *necessitating rule* with regard to α, since when ϕ is the case, it is also necessary that α is the case. Correspondingly, we call (3.6b) a *qualifying rule* with regard to β, because β may be the case only when ψ is also the case. By this definition, rule $O(\beta \rightarrow \alpha)$ is both necessitating with regard to α, and qualifying with regard to β. Furthermore, (3.6) entails that normalised rules do not contain additional applications of the deontic operator other than the enclosing one.

In both kinds of normalised rules, we refer to α and β as the *subject* of the rule, and to the complex terms ϕ and ψ as the *constraint* of the rule, respectively.

Normalised Rules as Hypotheses

Normalised rules with subjects in the form of *configured*(*o*) as in (3.4) express tailoring hypotheses about *o*. However, our notion of normalised rules is broader than that, and also comprises rules with other subjects in **A**.

Example 3.1 *Consider FPV*

$$small\text{-}team = proposition(\text{project, team-size}, ramp_{6,3}) \in \mathbf{A} \qquad (3.7)$$

defined as a fuzzy ramp where the "smallness" property degrades from 100% to 0% as the team size increases from 3 to 6 members. Then normalised rule

$$O(\neg configured(\text{weekly-team-meetings}) \to small\text{-}team) \qquad (3.8)$$

is a necessitating rule with regard to FPV small-team: If no weekly team meetings are being held, the team needs to be small.

For the purpose of finding an optimal tailoring configuration, we only need hypotheses about tailoring options in order to rate tailoring decisions. Hypotheses about properties of the tailoring context cannot contribute to find a better tailoring configuration and can thus be safely ignored.

However, hypotheses about the tailoring context can provide useful insights to the process tailorer. They can serve to rate the appropriateness of measurements in the tailoring context. By this the process tailorer can check whether the tailoring context specified by him is well-suited to a given tailoring configuration, or even whether there are contradictions within the tailoring context itself (see Example 3.2 below). We will discuss this further in Section 3.2 on justifications about tailoring decisions and the tailoring context.

For now, in order to handle hypotheses about arbitrary FPVs in **A**, we extend the domain of hypothesis functions as originally defined in (2.36) to

$$hyp_{n}, hyp_{q} : \mathbf{A} \mapsto \mathbf{F} \qquad (3.9)$$

and henceforth consider expressions of the form $hyp_{x}(o \in \mathbf{O})$ as shorthand notations for $hyp_{x}(configured(o))$.

Example 3.2 *From* (3.8) *we can derive the necessitating hypothesis*

$$hyp_{n}(small\text{-}team) = \neg configured(\text{weekly-team-meetings}) \qquad (3.10)$$

and, by transforming (3.8) *to its equivalent*

$$O(\neg small\text{-}team \to configured(\text{weekly-team-meetings})) \qquad (3.11)$$

we get

$$hyp_n(\text{weekly-team-meetings})$$
$$= hyp_n(\text{configured}(\text{weekly-team-meetings})) = \neg small\text{-}team \qquad (3.12)$$

Assume the process tailorer has specified a team size of 5 and has manually disabled weekly team meetings in the tailoring configuration. Then, next to issuing a low rating for option weekly-team-meetings, *the system could also alert the process tailorer that the team size conflicts with the current tailoring configuration because it does not satisfy the requirements for disabling option* weekly-team-meetings.

Axioms of Deontic Logic

Deontic logic commonly uses two additional auxiliary operators that can be defined in terms of O: $P\phi = \neg O\neg\phi$ denotes "it is permitted that ϕ is the case," and $F\phi = O\neg\phi$ denotes "it is forbidden that ϕ is the case."

Deontic logic shares two axioms with most other logics of the modal family:

$$\phi \rightarrow O\phi \qquad (3.13a)$$
$$(O(\phi \rightarrow \psi) \wedge O\phi) \rightarrow O\psi \qquad (3.13b)$$

meaning that what is the case is also obligatory (3.13a),[1] and that when it is obligatory that ϕ entails ψ, and in addition it is obligatory that ϕ is the case, then it is also obligatory that ψ is the case (3.13b).

There is also an additional axiom unique to deontic logic:

$$O(O\phi \rightarrow P\phi) \qquad (3.13c)$$

By this we demand that what is obligatory must also be permitted; and since $\neg P\phi$ is equivalent to $O\neg\phi$, this accommodates the intuition that it is not sensible to state that $O\phi$ and $O\neg\phi$ at the same time. The surrounding O operator concedes that, while it is desirable, it can not be guaranteed that a given set of rules actually contains no contradictions of the form $O\phi \wedge \neg P\phi$. For this same reason, the rating mechanism for tailoring decisions as described in Section 2.4.3 is designed to deal with the fact that a tailoring option is necessary, but not feasible, which is equivalent to stating that $O(\text{configured}(o)) \wedge \neg P(\text{configured}(o))$, and hence that $\neg(O(\text{configured}(o)) \rightarrow P(\text{configured}(o)))$.

We will also make use of the following corollary of axioms (3.13) [Garo5]:

$$O(\phi \wedge \psi) \equiv O\phi \wedge O\psi \qquad (3.14)$$

---

[1]This axiom is so self-evident it can lead to confusion. Antoine de Saint-Exupéry in his book *Le Petit Prince* [SE43] draws humour from this in his description of the king who believes he rules the stars but in fact orders them to do the things that they would naturally do anyway.

Avoiding Misinterpretations of Natural Language

We have chosen the wording of hypotheses (3.3) very carefully, and have used more words than may have been necessary in every-day language. The reason behind this is that intuitive translations from natural-language statements to statements in formal modal logic can lead to obscure errors or seeming contradictions.[2] Since the correct representation of natural-language statements in logic is rarely treated in depth in standard present-day literature, we now give an example of a common misinterpretation of common language. In (3.3a) we did not state, for instance, that

$$\text{if } \phi \text{ is the case, then so should } \psi \tag{3.15}$$

because it could too easily invoke the logic term

$$\phi \rightarrow O\,\psi \tag{3.16}$$

While this may resemble the syntactic structure of statement (3.15), there is a subtle difference in semantics: Operator O in (3.16) only changes the modality of the consequent ψ, while the modal verb "should" changes the modality of the whole statement. We can grasp this more easily when we rewrite (3.16) as the equivalent term

$$\neg\phi \vee O\,\psi \tag{3.17}$$

Clearly, (3.15) does not intend to express that either ϕ is not the case, or it is obligatory that ψ is the case. Rather, the intuition of (3.15) is that *it ought to be the case* that either ϕ is not case, or also ψ is, which amounts to the logic term

$$O(\neg\phi \vee \psi) \tag{3.18a}$$

or, equivalently, to

$$O(\phi \rightarrow \psi) \tag{3.18b}$$

The hypotheses as stated in (3.3) eliminate any ambiguity about this broader scope of the "obligation" modality.

3.1.2 General Tailoring Rules

We now have the formal foundations for discussing how tailoring rules can be provided to the TSS. The practical value of a TSS depends, among others, on low entry barriers and

---

[2]Mediæval scholastic logicians knew of these subtleties of every-day language that surface only when complex logical arguments are drawn up. Lacking today's formal methods, they developed a complex system of Latin parlance. It was later ridiculed as too cumbersome and discarded by Renaissance logicians, and is only being slowly re-discovered within the context of present-day formal logic [ØH95].

a steep initial learning curve for its users. A process modeler should be able to express his knowledge about software processes as flexibly and intuitively as possible. Some degree of abstract thinking will be indispensable, so what we do expect of the process modeler is the ability to express tailoring rules in terms of FFs $\phi \in \mathbf{F}$.

The tailoring framework outlined in Chapter 2 depends on tailoring hypotheses for individual tailoring options, as in the following example:

Example 3.3 (mutually exclusive tailoring options) *Suppose that the process modeler wants to forbid that tailoring options o_1 and o_2 both be chosen in the same tailoring configuration. He could express this condition as two qualifying hypotheses:*

$$hyp_q(o_1) = \neg configured(o_2) \qquad (3.19a)$$

$$hyp_q(o_2) = \neg configured(o_1) \qquad (3.19b)$$

stating that option o_1 is feasible only if option o_2 is not configured, and vice versa.

By having to define tailoring rules as tailoring hypotheses for individual tailoring options, the process modeler is forced to express every rule in terms of a specific tailoring option. In the case of Example 3.3, he is even forced to define two hypotheses in order to ensure the mutual exclusiveness of options o_1 and o_2, while he perceives this circumstance as one unique rule. Also, the process modeler might at some later time have to extend the hypothesis for option o_1 or o_2 in order to accommodate additional restrictions. As a consequence, the relationship between these two options expressed in Example 3.3 will become obscured.

The process modeler can express his knowledge about process tailoring more clearly and intuitively if we remove this restriction and allow him to express tailoring rules in terms of *general tailoring rules*. By this we understand arbitrary FFs in the form of $O\,\phi$.

Example 3.4 (general tailoring rule) *The hypotheses (3.19) from Example 3.3 can be restated as a single general tailoring rule*

$$O\,\neg(configured(o_1) \wedge configured(o_2)) \qquad (3.20)$$

The tailoring framework outlined in Chapter 2 cannot be applied to arbitrary rules such as (3.20). However, we will show in the following that any set of such arbitrary rules can be automatically transformed to an equivalent set of normalised rules, provided that for any rule $O\,\phi$, there is no further occurrence of O within ϕ. Since necessitating and qualifying tailoring hypotheses are just another syntactic variant of normalised rules, we can then apply the tailoring framework to all normalised rules of the form of (3.4).

Example 3.5 (normalised general tailoring rule) *The general tailoring rule (3.20) is equivalent to either of the two normalised rules*

$$O(configured(o_1) \rightarrow \neg configured(o_2)) \tag{3.21a}$$

$$O(configured(o_2) \rightarrow \neg configured(o_1)) \tag{3.21b}$$

which in turn can be restated as tailoring hypotheses (3.19): According to rule (3.21a), for $configured(o_1)$ to be true we require $configured(o_2)$ to be false. Qualifying hypothesis (3.19a) states the same in a different way—option o_1 only qualifies if $configured(o_2)$ is false. In the same way, rule (3.21b) corresponds to hypothesis (3.19b).

The process modeler does not need to know anything about deontic logic. He can express tailoring conditions as ordinary logic propositions $\phi \in F$, without the leading O operator. Assuming that every condition ϕ put forward by the process modeler "ought to be the case," we can apply the deontic operator implicitly. By not making the O operator available to the process modeler, we also make sure that ϕ does not contain additional applications of O—a condition we will rely on later on.

With the aid of deontic logic and the notion of normalised tailoring rules, we have shown the interrelations between arbitrary tailoring rules and tailoring hypotheses. In the following section, we will show how to systematically transform general tailoring rules to normalised rules, and in Section 3.1.4 we will restate normalised tailoring rules as necessitating and qualifying hypotheses.

3.1.3 Normalising General Tailoring Rules

A set of general tailoring rules can be systematically converted to an equivalent set of normalised tailoring rules. In this section, we will develop a method to accomplish this.

On the outset, we have a set of general tailoring rules

$$R = \{O\,\phi_1, \ldots, O\,\phi_n\} \tag{3.22}$$

Our goal is to obtain an equivalent set R^* of normalised tailoring rules in the form of (3.6), i.e.

$$R^* = \{O(\alpha_1 \rightarrow \psi_1), \ldots, O(\alpha_n \rightarrow \psi_n), O(\omega_1 \rightarrow \beta_1), \ldots, O(\omega_m \rightarrow \beta_m)\} \tag{3.23}$$

such that $R \equiv R^*$, which can then be translated to tailoring hypotheses about the subjects $\alpha_{1...n}$ and $\beta_{1...m}$ of these rules.

This transformation takes place in three steps. In step one we convert all conditions in R to *negation normal form* (NNF) in order to facilitate step two, where we actually transform every rule in R to one or more corresponding normalised rules. Finally, in step three, we simplify the normalised rules to make them more comprehensible and to make their evaluation more efficient.

Step One: Negation Normal Form

A logic term in NNF is composed only of atoms $\alpha \in \mathbf{A}$ and negated atoms $\neg\alpha$ that are connected with logic operators $\wedge$ and $\vee$. Apart from that, a NNF contains no other logic operators, and contains no negations of complex subterms.

Any term in $\mathbf{F}$ can be transformed into a NNF by function *nnf* defined recursively as follows:

$$nnf(\phi \to \psi) = nnf(\neg\phi) \vee nnf(\psi) \tag{3.24a}$$
$$nnf(\neg(\phi \to \psi)) = nnf(\phi) \wedge nnf(\neg\psi) \tag{3.24b}$$
$$nnf(\phi \wedge \psi) = nnf(\phi) \wedge nnf(\psi) \tag{3.24c}$$
$$nnf(\neg(\phi \wedge \psi)) = nnf(\neg\phi) \vee nnf(\neg\psi) \tag{3.24d}$$
$$nnf(\phi \vee \psi) = nnf(\phi) \vee nnf(\psi) \tag{3.24e}$$
$$nnf(\neg(\phi \vee \psi)) = nnf(\neg\phi) \wedge nnf(\neg\psi) \tag{3.24f}$$
$$nnf(\neg\neg\phi) = nnf(\phi) \tag{3.24g}$$
$$nnf(\alpha) = \alpha \tag{3.24h}$$
$$nnf(\neg\alpha) = \neg\alpha \tag{3.24i}$$

We further extend *nnf* to include simplifications such as the elimination of constants and duplicates:

$$nnf(\text{true} \wedge \phi) = nnf(\phi) \tag{3.24j}$$
$$nnf(\phi \wedge \text{true}) = nnf(\phi) \tag{3.24k}$$
$$nnf(\text{false} \wedge \phi) = nnf(\text{false}) \tag{3.24l}$$
$$nnf(\phi \wedge \text{false}) = nnf(\text{false}) \tag{3.24m}$$
$$nnf(\text{true} \vee \phi) = nnf(\text{true}) \tag{3.24n}$$
$$nnf(\phi \vee \text{true}) = nnf(\text{true}) \tag{3.24o}$$
$$nnf(\text{false} \vee \phi) = nnf(\phi) \tag{3.24p}$$
$$nnf(\phi \vee \text{false}) = nnf(\phi) \tag{3.24q}$$
$$nnf(\phi \wedge \phi) = nnf(\phi) \tag{3.24r}$$
$$nnf(\phi \vee \phi) = nnf(\phi) \tag{3.24s}$$

In addition to supporting the standard operators of propositional calculus, we also allow function *nnf* to transform some additional short-hand notations that make it easier for the process modeler to define tailoring conditions:

$$nnf(\textit{if-then}(\phi, \psi)) = nnf(\phi \rightarrow \psi) \tag{3.24t}$$

$$nnf(\textit{if-then-else}(\phi, \psi, \omega)) = nnf((\phi \rightarrow \psi) \wedge (\neg\phi \rightarrow \omega)) \tag{3.24u}$$

$$nnf(\textit{mutex}(\phi_1, \ldots, \phi_n)) = nnf\left(\bigwedge_{1 \leq i \leq n} \left(\phi_i \rightarrow \neg\bigvee_{j \in \{1\ldots n\} \backslash i} \phi_j\right)\right) \tag{3.24v}$$

$$nnf(\textit{mutual}(\phi_1, \ldots, \phi_n)) = nnf\left(\left(\bigvee_{1 \leq i \leq n} \phi_i\right) \rightarrow \left(\bigwedge_{1 \leq i \leq n} \phi_i\right)\right) \tag{3.24w}$$

if-then is just another way for stating an implication, *if-then-else* allows to state an alternative for the case that the antecedent does not hold. *mutex* is an exclusive *or*: It holds if and only if exactly one of its operands hold. *mutual* holds either if none of its operands hold, or if all of its operands hold.

Function *nnf* maintains equivalence for all terms $\phi \in \mathbf{F}$, i.e.

$$nnf(\phi) \equiv \phi \tag{3.25}$$

since applications (3.24a)–(3.24s) of function *nnf* reflect standard transformations of propositional calculus [W$^+$o5b], and applications (3.24t)–(3.24w) maintain equivalence by definition.

Example 3.6 (negation normal form) *Function nnf normalises logic term*

$$mutual(\alpha, \beta) \tag{3.26}$$

as follows:

$$nnf(\textit{mutual}(\alpha, \beta)) \tag{3.27a}$$

$$= nnf((\alpha \vee \beta) \rightarrow (\alpha \wedge \beta)) \tag{3.27b}$$

$$= nnf(\neg(\alpha \vee \beta)) \vee nnf(\alpha \wedge \beta) \tag{3.27c}$$

$$= nnf(\neg\alpha \wedge \neg\beta) \vee nnf(\alpha) \wedge nnf(\beta) \tag{3.27d}$$

$$= nnf(\neg\alpha) \wedge nnf(\neg\beta) \vee \alpha \wedge \beta \tag{3.27e}$$

$$= \neg\alpha \wedge \neg\beta \vee \alpha \wedge \beta \tag{3.27f}$$

As expected, all negations in (3.27f) are applied to atomic subterms only, and the only operands used are $\wedge$ and $\vee$.

Step Two: Rule Normalisation

We first transform R to R' as follows:

$$R' = \bigcup_{O(\phi) \in R} O(\text{true} \to \mathit{nnf}(\phi)) \qquad (3.28)$$

Then we recursively normalise R' to obtain R^*:

$$R^* = \mathit{normalise}(R') \qquad (3.29)$$

with function $\mathit{normalise} : \mathcal{P}(\mathbf{F}) \mapsto \mathcal{P}(\mathbf{F})$ defined as

$$\mathit{normalise}(\{O(\phi \to (\psi \wedge \omega))\} \cup \Gamma) = \mathit{normalise}(\{O(\phi \to \psi), O(\phi \to \omega)\} \cup \Gamma) \qquad (3.30a)$$
$$\mathit{normalise}(\{O(\phi \to (\psi \vee \omega))\} \cup \Gamma) = \mathit{normalise}($$
$$\{O((\phi \wedge \neg\psi) \to \omega), O((\phi \wedge \neg\omega) \to \psi)\} \cup \Gamma) \qquad$$
$$(3.30b)$$
$$\mathit{normalise}(\{O(\phi \to \neg\alpha)\} \cup \Gamma) = \{O(\mathit{simplify}(\alpha \to \neg\phi))\} \cup \mathit{normalise}(\Gamma) \qquad (3.30c)$$
$$\mathit{normalise}(\{O(\phi \to \alpha)\} \cup \Gamma) = \{O(\mathit{simplify}(\phi \to \alpha))\} \cup \mathit{normalise}(\Gamma) \qquad (3.30d)$$
$$\mathit{normalise}(\emptyset) = \emptyset \qquad (3.30e)$$

with $\Gamma \subseteq \mathbf{F}$.

Since function *normalise* operates only on implications with consequents in NNF, definition (3.30) is exhaustive.

The idea of rule normalisation can be described as follows in a procedural fashion: We start out with a set R' as specified in (3.28). The condition of every rule is an implication whose antecedent is trivial and whose consequent is a NNF. We now pick a rule $O(\phi \to \psi)$ from R' and remove it from R'. If its consequent ψ is a conjunction or a disjunction, we create two new rules as specified by (3.30a) and (3.30b) and add them back to R'. Otherwise, the consequent ψ is an atomic term from $\mathbf{A}$, or the negation of an atomic term. If we have a negated atomic term, we get a qualifying rule by reversing the direction of the implication and add it to R^* (3.30c). With an atomic term as consequent, we have a necessitating rule which we can add to R^* without further modifications (3.30d). To render rules as compact and comprehensible as possible, (3.30c) and (3.30d) simplify the conditions of both necessitating and qualifying rules. We will deal with this simplification step separately further below. The transformation continues until there are no more rules in R' left to be transformed (3.30e).

In a manner of speaking, we use the implication operator '$\to$' as a pivot, gradually moving parts of complex consequents ψ onto the side of the antecedent ϕ, until only an atomic term, or the negation of an atomic term, remains as the consequent.

Example 3.7 *Consider a singleton rule set that incorporates the mutuality condition (3.26) from Example 3.6:*

$$R = \{O(mutual(\alpha, \beta))\} \qquad (3.31)$$

With the normalised condition (3.27f) we get

$$R' = \{true \rightarrow (\neg\alpha \wedge \neg\beta \vee \alpha \wedge \beta)\} \qquad (3.32)$$

and calculate the set of normalised rules as follows:

$$R^* = normalise(\{ \qquad (3.33a)$$
$$O(true \rightarrow (\neg\alpha \wedge \neg\beta \vee \alpha \wedge \beta))\}) \qquad (3.33b)$$
$$= normalise(\{$$
$$O((true \wedge \neg(\neg\alpha \wedge \neg\beta)) \rightarrow (\alpha \wedge \beta)),$$
$$O((true \wedge \neg(\alpha \wedge \beta)) \rightarrow (\neg\alpha \wedge \neg\beta))\}) \qquad (3.33c)$$
$$= normalise(\{$$
$$O((true \wedge \neg(\neg\alpha \wedge \neg\beta)) \rightarrow \alpha),$$
$$O((true \wedge \neg(\neg\alpha \wedge \neg\beta)) \rightarrow \beta),$$
$$O((true \wedge \neg(\alpha \wedge \beta)) \rightarrow \neg\alpha),$$
$$O((true \wedge \neg(\alpha \wedge \beta)) \rightarrow \neg\beta)\}) \qquad (3.33d)$$
$$= \{O(simplify(true \wedge \neg(\neg\alpha \wedge \neg\beta) \rightarrow \alpha)),$$
$$O(simplify(true \wedge \neg(\neg\alpha \wedge \neg\beta) \rightarrow \beta)),$$
$$O(simplify(\alpha \rightarrow \neg(true \wedge \neg(\alpha \wedge \beta)))),$$
$$O(simplify(\beta \rightarrow \neg(true \wedge \neg(\alpha \wedge \beta))))\} \qquad (3.33e)$$

As is to be expected, the resulting set of rules (3.33e) only contains normalised rules. Also note the impact of applying function nnf in the last step of rule normalisation. It can be verified intuitively that (3.33e), in terms of necessitating and qualifying rules, expresses the same as the initial rule put forward in (3.31): That neither α nor β may occur in isolation.

Step Three: Simplification

We now proceed to define function *simplify* that performs the third and final step of rule normalisation—simplifying the constraints of normalised necessitating and qualifying rules:

$$simplify(\phi \rightarrow \alpha) = nnf(\phi[\alpha \curvearrowright false]) \rightarrow \alpha \qquad (3.34a)$$
$$simplify(\alpha \rightarrow \phi) = \alpha \rightarrow nnf(\phi[\alpha \curvearrowright true]) \qquad (3.34b)$$

where $\phi[\rho \curvearrowright \sigma]$ is defined as the *substitution* operation that substitutes σ for all occurrences of ρ in term ϕ.

With function *simplify* we achieve two things: First, we eliminate self-references from the constraints ϕ of normalised rules to their respective subjects α by substituting logic constants for all occurrences of that subject α within ϕ (see (3.37c) in Example 3.8 below). Second, we replace the resulting constraints with their NNFs in order to eliminate unnecessary complexity introduced by rule normalisation and term substitution.

To justify that the substitutions in (3.34) are admissible, we need to show that they maintain equivalence, i.e.

$$\phi \to \alpha \equiv \phi[\alpha \curvearrowright \text{false}] \to \alpha \tag{3.35a}$$

$$\alpha \to \phi \equiv \alpha \to \phi[\alpha \curvearrowright \text{true}] \tag{3.35b}$$

Let us first consider substitutions in necessitating rules (3.35a). If α is the case, then any implication $\psi \to \alpha$ holds, therefore equivalence (3.35a) holds. If α is not the case, then it is permissible to substitute 'false' for α, so again equivalence (3.35a) holds.

We argue similarly for substitutions in qualifying rules (3.35b): If α is the case, then it is permissible to substitute 'true' for α, therefore equivalence (3.35b) holds. If α is not the case, then any implication $\alpha \to \phi$ holds, so again equivalence (3.35b) holds.

Hence, substitutions (3.35) maintain equivalence. Since function *nnf*, too, maintains equivalence (3.25), we can conclude that

$$simplify(\phi) \equiv \phi \tag{3.36}$$

Example 3.8 (Simplifying normalised rules) *With the above definition of function simplify, we can continue the evaluation of (3.33e) as follows:*

$$
\begin{aligned}
R^* = \{ &\text{O}(simplify(\text{true} \wedge \neg(\neg\alpha \wedge \neg\beta) \to \alpha)), \\
&\text{O}(simplify(\text{true} \wedge \neg(\neg\alpha \wedge \neg\beta) \to \beta)), \\
&\text{O}(simplify(\alpha \to \neg(\text{true} \wedge \neg(\alpha \wedge \beta)))), \\
&\text{O}(simplify(\beta \to \neg(\text{true} \wedge \neg(\alpha \wedge \beta))))\} \\
\end{aligned}
\tag{3.37a}
$$

$$
\begin{aligned}
= \{ &\text{O}(nnf((\text{true} \wedge \neg(\neg\alpha \wedge \neg\beta))[\alpha \curvearrowright \text{false}]) \to \alpha), \\
&\text{O}(nnf((\text{true} \wedge \neg(\neg\alpha \wedge \neg\beta))[\beta \curvearrowright \text{false}]) \to \beta), \\
&\text{O}(\alpha \to nnf((\neg(\text{true} \wedge \neg(\alpha \wedge \beta)))[\alpha \curvearrowright \text{true}])), \\
&\text{O}(\beta \to nnf((\neg(\text{true} \wedge \neg(\alpha \wedge \beta)))[\beta \curvearrowright \text{true}]))\} \\
\end{aligned}
\tag{3.37b}
$$

$$
\begin{aligned}
= \{ &\text{O}(nnf(\text{true} \wedge \neg(\neg\text{false} \wedge \neg\beta)) \to \alpha), \\
&\text{O}(nnf(\text{true} \wedge \neg(\neg\alpha \wedge \neg\text{false})) \to \beta), \\
&\text{O}(\alpha \to nnf(\neg(\text{true} \wedge \neg(\text{true} \wedge \beta)))), \\
\end{aligned}
$$

$$O(\beta \rightarrow nnf(\neg(\text{true} \wedge \neg(\alpha \wedge \beta))))\} \qquad (3.37\text{c})$$
$$= \{O(\beta \rightarrow \alpha),$$
$$O(\alpha \rightarrow \beta)\} \qquad (3.37\text{d})$$

There are only two rules left in (3.37d) *because applying function nnf in* (3.37c) *leaves us with two pairs of duplicate rules.*

Equivalence of Rules and Normalised Rules

We now provide a justification that R and R^* are equivalent.

R and R' are equivalent since for each rule $O\phi \in R$ we have a corresponding rule $O(\text{true} \rightarrow nnf(\phi)) \in R'$, and we have

$$\phi \equiv nnf(\phi) \equiv \text{true} \rightarrow nnf(\phi) \qquad (3.38)$$

We can justify that R' and R^* are equivalent by showing that transformations (3.30) each result in equivalent sets of rules. With the axioms of standard propositional calculus [W$^+$05b], extended by the axioms of deontic logic (3.13), we have:

$$\{O(\phi \rightarrow (\psi \wedge \omega))\} \equiv \{O(\phi \rightarrow \psi \wedge \phi \rightarrow \omega)\}$$
$$\equiv \{O(\phi \rightarrow \psi) \wedge O(\phi \rightarrow \omega)\}$$
$$\equiv \{O(\phi \rightarrow \psi), O(\phi \rightarrow \omega)\} \qquad (3.39\text{a})$$
$$\{O(\phi \rightarrow (\psi \vee \omega))\} \equiv \{O((\phi \wedge \neg\psi) \rightarrow \omega \wedge (\phi \wedge \neg\omega) \rightarrow \psi)\}$$
$$\equiv \{O((\phi \wedge \neg\psi) \rightarrow \omega), O((\phi \wedge \neg\omega) \rightarrow \psi)\} \qquad (3.39\text{b})$$

Therefore (3.30a) and (3.30b) result in equivalent sets of rules. Since $\phi \rightarrow \psi \equiv \neg\psi \rightarrow \neg\phi$ and *simplify*$(\phi) \equiv \phi$ (3.36), this also applies for (3.30c) and (3.30d). Finally, (3.30e) does not rewrite rules and consequently also maintains equivalence of R' and R^*.

Hence, $R \equiv R'$ and $R' \equiv R^*$, and therefore $R \equiv R^*$.

3.1.4 Translating Normalised Rules to Tailoring Hypotheses

Given a set of normalised rules R^*, we can now set about deriving necessitating and qualifying hypotheses $hyp_n(\alpha)$ and $hyp_q(\alpha)$ for every atomic term $\alpha \in \mathbf{A}$ with necessitating rules $O(\phi \rightarrow \alpha)$ and qualifying rules $O(\alpha \rightarrow \phi)$ in R^*, as for atoms α and β in the following example:

Example 3.9 *Given* $R^* = \{O(\alpha \rightarrow \beta), O(\alpha \rightarrow \phi), O(\beta \rightarrow \phi)\}$ *we have two qualifying rules* $O(\alpha \rightarrow \beta)$ *and* $O(\alpha \rightarrow \phi)$ *for* α, *one qualifying rule* $O(\beta \rightarrow \phi)$ *for* β, *and one necessitating rule* $O(\alpha \rightarrow \beta)$ *for* β.

As Example 3.9 illustrates, the number of qualifying or necessitating rules can vary for each atom in **A**: There may be no rule at all, one rule, or more than one rule. In the case of more than one necessitating or qualifying rule, we need to collate all rules into a single logic term for the according hypothesis.

Before we detail the collation procedure, we need to be aware of a basic assumption about the completeness of the rules provided.

Closed World Assumption

We consider the rules provided by the process modeler in R, and therefore the rules in R^*, to be *complete*. By this we mean that if for any ϕ its necessity ($O\phi$) or unfeasibility ($O\neg\phi = F\alpha$) cannot be inferred by means of the given rules, we can safely assume that it is in fact not necessary or unfeasible, respectively. We are thus dealing with a *closed world* with regard to rules, where any rule $O\phi$ that cannot be explicitly proven to exist can be considered not to exist, i.e., our default assumption is $\neg O\phi$, meaning that ϕ is by default not necessary. Accordingly, unless otherwise stated, we assume by default that $\neg O\neg\psi$, meaning that it is not necessary that $\neg\psi$, or, equivalently, that it is feasible that ϕ ($\neg O\neg\psi = P\psi$).

This is in opposition to an *open world* assumption where the lack of evidence does not by itself entail the absence of a rule. Were we instead to assume an open world, we would never be able to say with certainty of any $\alpha \in \mathbf{A}$ that it was really unnecessary, or feasible, since a rule unknown to us could account for the opposite.

Rule Collation

We collate necessitating hypotheses about $\alpha \in \mathbf{A}$ as

$$hyp_n(\alpha) = \bigvee_{O(\phi \rightarrow \alpha) \in R^*} \phi \qquad (3.40)$$

This can be justified as follows: We can replace all necessitating rules $O(\phi_1 \rightarrow \alpha), \ldots, O(\phi_n \rightarrow \alpha)$ in R^* by a singular term $O(\phi_1 \rightarrow \alpha) \wedge \ldots \wedge O(\phi_n \rightarrow \alpha)$ each, and then, due to the distributivity of operators O and $\wedge$ [Garo5], by rule $O((\phi_1 \rightarrow \alpha) \wedge \ldots \wedge (\phi_n \rightarrow \alpha))$. Propositional calculus allows us to rewrite this as $O((\phi_1 \vee \ldots \vee \phi_n) \rightarrow \alpha)$, which is in turn a necessitating rule. The resulting necessitating hypothesis is $hyp_n(\alpha) = \phi_1 \vee \ldots \vee \phi_n$.

With only one necessitating rule for α we have the trivial case $hyp_n(\alpha) = \phi_1$. If no necessitating rule is given for α, we get $hyp_n(\alpha) = \bigvee_\emptyset \phi = $ false (cf. (3.2c)), which is in line with our closed world assumption.

Similarly, we collate qualifying hypotheses about α as

$$hyp_q(\alpha) = \bigwedge_{O(\alpha \to \phi) \in R^*} \phi \tag{3.41}$$

with an analogous justification: We combine multiple qualifying rules $O(\alpha \to \phi_1), \ldots, O(\alpha \to \phi_n)$ to $O((\alpha \to \phi_1) \wedge \ldots \wedge (\alpha \to \phi_n))$ and get the unified qualifying rule $O(\alpha \to (\phi_1 \wedge \ldots \wedge \phi_n))$ and qualifying hypothesis $hyp_q(\alpha) = \phi_1 \wedge \ldots \wedge \phi_n$. With $n = 1$ we have $hyp_q(\alpha) = \phi_1$. If no qualifying rule is given for α, then α can never be disqualified due to the closed world assumption, so with $n = 0$ we have $hyp_q(\alpha) = \bigwedge_\emptyset \phi = $ true (cf. (3.2d)).

3.1.5 Additional Properties of Tailoring Rule Normalisation

Handling Contradictions

With deontic logic we can manage logic contradictions—both within single rules and across different rules. A logic contradiction is any equivalent of the statement $\phi \wedge \neg\phi$. In any system of logic propositions, the occurrence of one of these contradictions causes a phenomenon known as *ex falso quodlibet* [Sha05]: Anything can be deduced from such a contradictory system.

By virtue of our use of the deontic operator O, we gracefully evade this condition. After all, we are not stating the impossible, we are only asking it, which in itself is not a contradiction. This can best be illustrated by an example:

Example 3.10 (internal contradiction) *Suppose that we have a only a single rule:*

$$R = \{O(configured(o) \wedge \neg configured(o))\} \tag{3.42}$$

From this we get

$$R' = \{O(true \to (configured(o) \wedge \neg configured(o)))\} \tag{3.43}$$

and (with false *substituted for* $\neg$true*)*

$$R^* = \{O(true \to configured(o)), O(configured(o) \to false)\} \tag{3.44}$$

We can restate R^* as the tailoring hypotheses

$$hyp_n(o) = true \tag{3.45a}$$
$$hyp_q(o) = false \tag{3.45b}$$

Example 3.10 results in a necessitating hypothesis demanding that o should always be configured, and a qualifying hypothesis stating that it is never feasible to configure o. In Section 2.4.3 we have described this situation as a *clash* and have shown that the rating mechanism handles such contradictions gracefully. It will assign a low rating regardless of whether option o is decided for, or against.

Another interesting case is two different rules contradicting each other:

Example 3.11 (two contradicting rules) *Suppose that we have two rules:*

$$R = \{O(\mathit{configured}(o_1) \rightarrow \mathit{configured}(o_2)), O(\mathit{configured}(o_1) \rightarrow \neg\mathit{configured}(o_2))\} \tag{3.46}$$

The contradiction is evident: For the case that o_1 is configured, we once demand that so should o_2, and at the same time demand that in the same case o_2 should not be configured. From R we obtain by normalisation

$$\begin{aligned} R^* = \{ & O(\mathit{configured}(o_1) \rightarrow \mathit{configured}(o_2)), \\ & O(\mathit{configured}(o_1) \rightarrow \neg\mathit{configured}(o_2)) \\ & O(\mathit{configured}(o_2) \rightarrow \neg\mathit{configured}(o_1))\} \end{aligned} \tag{3.47}$$

From this we get the tailoring hypotheses

$$\mathit{hyp}_n(o_1) = \text{false} \tag{3.48a}$$
$$\mathit{hyp}_q(o_1) = \mathit{configured}(o_2) \wedge \neg\mathit{configured}(o_2) \tag{3.48b}$$
$$\mathit{hyp}_n(o_2) = \mathit{configured}(o_1) \tag{3.48c}$$
$$\mathit{hyp}_q(o_2) = \neg\mathit{configured}(o_1) \tag{3.48d}$$

Example 3.11 illustrates how contradictions across rules show up in the hypotheses for all tailoring options involved. The qualifying hypothesis for option o_1 will never yield a rating of 1. Rating option o_2 will result in a clash in the same fashion as discussed for Example 3.10 above.

Redundancy

Rule normalisation introduces redundant rules. Specifically, case (3.30b) of function *normalise* replaces one original rule $O(\phi \rightarrow (\psi \vee \omega))$ with two mutually equivalent rules, $O((\phi \wedge \neg\psi) \rightarrow \omega)$ and $O((\phi \wedge \neg\omega) \rightarrow \psi)$. For the sake of obtaining an equivalent set of normalised rules R^* for a set of general tailoring rules R, it would be sufficient to keep only one of these two rules resulting from (3.30b).

However, by maintaining this redundancy, we make sure that any rule $O\,\phi$ is transformed to at least one normalised rule for every distinct atom in $O\,\phi$: (3.30a) and (3.30b) ensure that all complex subterms of ϕ are divided up into their constituents recursively and that, ultimately, for every atomic subterm α a rule will be generated in the form of $O(\psi \rightarrow \alpha)$ or $O(\omega \rightarrow \neg\alpha)$. As a consequence, we ensure that rule $O\,\phi$ will directly affect the ratings of all tailoring decisions and property valuations represented by the atomic terms in ϕ.

Furthermore, in Section 3.2 this redundancy will allow us to develop a mechanism for providing justifications for all individual ratings of both tailoring decisions and property valuations. These justifications will allow us to trace each rating back to all contributing facts and conditions.

3.2 Justifying Ratings of Tailoring Decisions

The tailoring framework put forward in Chapter 2 is complex and multi-layered: The TSS recommends a tailoring configuration (Section 2.5), which by definition is a collection of tailoring decisions, mapping all available tailoring options to one of the two states *chosen* or *excluded*. For every tailoring decision the TSS provides a pessimistic and an optimistic rating (Section 2.4.5). The TSS recommends tailoring decisions based on their optimistic ratings (Section 2.5.3), with the goal of optimising the overall optimistic rating (Section 2.4.4) of the tailoring configuration.

To understand a tailoring decision within the recommended tailoring configuration, the rating of the decision must be explained in terms of the tailoring hypotheses that were applied to calculate the rating. Decision ratings are calculated in two steps: For each tailoring option o, the TSS first calculates the valuations of the necessitating and qualifying hypotheses for o (Section 2.4.2), and subsequently combines these to calculate a pessimistic and an optimistic rating for the tailoring decision about o (see (2.46) in Section 2.4.5). Each hypothesis is represented by a logic term and is valuated in FIL. The atomic fuzzy formulae from which these terms are constructed refer either to decisions about other tailoring options, or to measurements from the tailoring context via fuzzy mappings.

This structure is too complex to be understood by a process tailorer without good knowledge of the internal workings of the TSS. Still, the process tailorer will expect to obtain an intuitive justification of tailoring decision ratings to understand the premises under which the TSS has made its recommendations. This allows the process tailorer to assess whether the TSS has ignored a relevant fact known to the process tailorer, or whether a fact the TSS considers relevant can actually be ignored under the current circumstances. By providing transparency for its ratings through appropriate justifications, the TSS increases the process tailorer's trust in its recommendations, and thus improves its acceptance and usefulness.

$$A_1 \xrightarrow{\text{because}} \begin{cases} A_{1.1} \\ A_{1.2} \xrightarrow{\text{because}} \begin{cases} A_{1.2.1} \xrightarrow{\text{because}} A_{1.2.1.1} \\ A_{1.2.2} \end{cases} \\ A_{1.3} \xrightarrow{\text{because}} A_{1.3.1} \end{cases}$$

Figure 3.1: A justification hierarchy

Our goal in this section therefore is to provide a uniform structure for justifications of the tailoring decisions put forward by the TSS, and a method for generating justifications according to that structure. First, we provide a formal structure for justifications in Section 3.2.1. Then we show how to justify the valuations of logic propositions in Section 3.2.2. After that, in Section 3.2.3, we construct justifications for tailoring decision ratings that build on the justifications for the underlying tailoring hypotheses.

Once we have established a method for justifying ratings of tailoring decisions, we can also derive a similar method to provide more insight into measurements from the tailoring context: We can use the same justification structure to explain why they have a positive or negative impact on ratings of one or more tailoring decisions. This allows for a quick review of whether and how adjustments to the tailoring context, such as augmenting the team by an additional member, or increasing the intended duration of a project, can best improve the quality of the process being tailored. We will discuss that in Section 3.2.4.

Even though the following sections will show that constructing a justification is in itself a complex matter, the result must be easily comprehensible. A proposal for visualising justifications will therefore conclude this discussion of justifications in Section 3.2.5.

3.2.1 Justifications

By the term *justification* we understand a claim supported by a collection of assertions, where the claim itself, too, is expressed in the form of an assertion. Each assertion within the justification can in turn be considered a claim in need of justification. With the term *justification hierarchy* we refer to a tree-like structure of assertions where the relation between each parent x and child y has the meaning "x because y" (Figure 3.1). Assertions that are not themselves justified take the role of axioms, i.e., generally accepted presumptions. Every finite justification hierarchy must ultimately recur to axioms.

Our goal is to justify ratings of tailoring decisions on the grounds of valuations of hypotheses. We therefore need to adapt the notion of assertions to our purpose, and accompany assertions with a pessimistic and an optimistic indication of their applicability, on a scale ranging from 0 (not applicable) to 1 (applicable), just as with fuzzy logic.

In the case that a tailoring option o has been chosen in the current tailoring configuration, the assertion is "tailoring option o should be chosen," and the interval of its pessimistic and optimistic ratings will express the applicability of that statement. We use the wording "should be chosen" instead of "is chosen" because the claim we want to justify is not the fact that it *has* been chosen, but why the TSS has declared, by a high rating, that the choice has been a good one—or a bad one in the case of a low decision rating. If a tailoring option o has been excluded from the tailoring configuration, the claim to be justified is "tailoring option o should be excluded."

The applicability of each of these two kinds of statements is equivalent to the FTI rating of the respective tailoring decisions.

As a formal representation of assertions and their justifications, we now introduce the set J of justification hierarchies. A justification hierarchy is a triple

$$(s, \hat{v}, J) \in J = S \times \hat{Q} \times \mathcal{P}(J) \tag{3.49}$$

where s is from the set of arbitrary natural language statements S, $\hat{v}$ is a FTI, and J is a set of justification hierarchies. Every justification hierarchy in J represents the assertion that the applicability of statement s is $\hat{v}$, along with one or more justifications J for that assertion.

A justification hierarchy represents an *axiom* when it consists of an assertion with an empty justification:

$$J_{\text{axiom}} = (S \times \hat{Q} \times \{\emptyset\}) \subset J \tag{3.50}$$

Example 3.12 *The statement*

(3.51) My clothes are wet because water is falling from the sky and I do not have an umbrella. Water is falling from the sky because it is raining.

can be formally represented as the justification hierarchy

$$\begin{aligned} (\text{``My clothes are wet''}, [1, 1], \{ \\ (\text{``water falling from sky''}, [1, 1], \{(\text{``it is raining''}, [1, 1], \emptyset)\}), \\ (\text{``I have an umbrella''}, [0, 0], \emptyset)\}) \in J \end{aligned} \tag{3.52}$$

where FTI $[1, 1]$ indicates that the associated statement is uncompromisingly true, whereas FTI $[0, 0]$ signifies that the associated statement is uncompromisingly false.

3.2.2 Justifying Valuations of Logic Propositions

We will now provide a scheme to justify valuations of logic propositions. We will develop this scheme in three steps: First, we will show how to justify valuations in classical (binary) logic. We will then extend the scheme to fuzzy valuations, and then to fuzzy interval valuations.

Justifications in Binary Logic

We justify the valuation of a logic term because we want to give an intuitive understanding of its valuation. The nature of the justification should therefore be as close to "common sense" as possible. To concretise our notion of common sense, we start off with an example:

Example 3.13 (justification) *Consider the following rule:*

(3.53)   IF *[you are about to go outside]$_\alpha$, and*
            [is is raining]$_\beta$ or no [sunny weather has been announced]$_\gamma$
            THEN *put your coat on.*

We can paraphrase the necessitating condition in (3.53), i. e., the "if" portion, as the logic proposition ϕ defined as

$$\phi = \alpha \wedge (\beta \vee \neg\gamma) \tag{3.54}$$

When given a recommendation such as

(3.55) *Put your coat on!*

we can ask "Why?" to obtain a justification. We will expect an answer which provides us with all relevant facts that contribute to the truth of condition (3.54). Suppose that we are about to go outside ($v(\alpha) =$ true), it is raining ($v(\beta) =$ true), and sunny weather has been announced ($v(\gamma) =$ true). With (3.54), we get $v(\phi) =$ true. We can then justify the truth of condition (3.54), and therefore the applicability of recommendation (3.55), as follows:

(3.56) *"Because [you are about to go outside]$_\alpha$ and [it is raining]$_\beta$."*

We can see immediately that the fact that sunny weather has been announced is not among the facts that could convince us to put on a coat, and hence is not relevant in the explanation. Note also that the justification does not reveal the structure of the underlying rule, instead it only provides a flat list of facts.

To justify the valuation of a logic proposition ϕ, our goal is thus to provide a list of facts we consider *relevant* to the valuation of ϕ. With Example 3.13 in mind, we will now provide a formal definition of relevant facts for justifying the valuation of a logic term.

Suppose we have a logic proposition $\phi \in \mathbf{F}$ that is written in NNF. This means that ϕ is composed of nested logical conjunctions and disjunctions, and negations are at most applied to atoms. Given a valuation function $v : \mathbf{F} \mapsto \mathbb{B}$, we call an atom $\alpha \in \mathbf{A}$ *relevant* to the overall valuation of ϕ if either:

1. α is a subterm of ϕ which is not negated, and $v(\alpha) = v(\phi)$, or

2. $\neg\alpha$ is a subterm of ϕ, and $v(\neg\alpha) = v(\phi)$.

This definition is based on the following intuition: If a conjunction $\phi_1 \wedge \ldots \wedge \phi_n$ or a disjunction $\phi_1 \vee \ldots \vee \phi_n$ valuate to true, then all subterms ϕ_i contribute to the overall valuation that themselves valuate to true. The same is the case if the conjunction or disjunction valuates to false: Equally, we consider all subterms ϕ_i contributors to the overall valuation that themselves valuate to false. If we apply this concept recursively to terms ψ expressed in NNF, we end up with atoms and negated atoms, each of which we consider to contribute to the truth or falseness of ψ if and only if their individual valuations are identical with the valuation of ψ.

Example 3.14 (relevant atoms) *Consider term*

$$\phi = \alpha \wedge (\neg\beta \vee \gamma) \tag{3.57}$$

with $\alpha, \beta, \gamma \in \mathbf{A}$ and $v(\alpha) =$ true and $v(\beta) = v(\gamma) =$ false. From this follows that $v(\phi) =$ true. Our above definition of relevance applies to atoms α and β: $v(\alpha) = v(\phi)$ and $v(\neg\beta) = v(\phi)$, whereas $v(\gamma) \neq v(\phi)$. We therefore justify the truth of ϕ by stating that α and $\neg\beta$.

If, however, we have $v(\alpha) =$ false, we also get $v(\phi) =$ false. We then justify the valuation of ϕ by stating that neither α nor γ are the case. In this case β is not relevant since $v(\neg\beta) \neq v(\phi)$.

With the above definition of relevant atoms, we can now recursively determine the set $L_{\phi,v}$ of relevant atoms in ϕ for valuation function v as follows:

$$L_{\phi,v} = relevant(nnf(\phi)) \tag{3.58}$$

where function *relevant* is defined as

$$relevant : \mathbf{F} \mapsto \mathcal{P}(\mathbf{A}) \tag{3.59a}$$

with

$$relevant(\psi \wedge \omega) = \begin{cases} relevant(\psi) & \text{if } v(\omega) \neq v(\phi) \\ relevant(\omega) & \text{if } v(\psi) \neq v(\phi) \\ relevant(\psi) \cup relevant(\omega) & \text{otherwise} \end{cases} \quad (3.59b)$$

$$relevant(\psi \vee \omega) = relevant(\psi \wedge \omega) \quad (3.59c)$$

$$relevant(\alpha) = \begin{cases} \{\alpha\} & \text{if } v(\alpha) = v(\phi) \\ \emptyset & \text{otherwise} \end{cases} \quad (3.59d)$$

$$relevant(\neg\alpha) = \begin{cases} \{\alpha\} & \text{if } v(\neg\alpha) = v(\phi) \\ \emptyset & \text{otherwise} \end{cases} \quad (3.59e)$$

with $\alpha \in \mathbf{A}$ and $\psi, \omega \in \mathbf{F}$.

Using the NNF of terms ϕ (3.58) to determine relevant atoms saves us from defining special cases for negated complex sub-terms of ϕ. The simplifications incorporated in our definition of function *nnf* also ensure that we do not get more explanations than necessary.

We can now go on to extend this concept to accommodate fuzzy logic.

Justifications in Fuzzy Logic

Fuzzy logic replaces categories *true* and *false* of binary logic with a continuous interval of *fuzzy truth values* (FTVs) ranging from 0 to 1 (Section 2.3.3). A simple example illustrates that function *relevant* as defined in (3.59) no longer yields desired results:

Example 3.15 (strict relevance with fuzzy valuations) *Given FF $\phi = \alpha \vee \beta$ and fuzzy valuations $v(\alpha) = 0.94$ and $v(\beta) = 0.95$, we get $v(\phi) = \max(v(\alpha), v(\beta)) = 0.95$. When calculating the set of relevant atoms for this valuation, we get relevant$(\alpha \vee \beta) =$ relevant$(\beta) = \{\beta\}$. α does not qualify as a relevant atom since $v(\alpha) \neq v(\phi)$.*

Example 3.15 shows that *relevant* is too strict for justifying fuzzy valuations. Intuitively, we would expect α to be part of the overall valuation because its valuation is very close to the overall valuation. But function *relevant* only selects sub-terms whose relevance is exactly equal to the valuation of the full term; given $v(\phi) = 0.95$, it does not make a difference whether we have $v(\alpha) = 0$ or $v(\alpha) = 0.94$: in both cases, α is not recognised as relevant.

To work around this problem, we map every FTV x to one of five disjunct truth categories:

$$category(x \in \mathbf{Q}) = \begin{cases} \text{false} & \text{if } x \leq m \\ \text{tendentially-false} & \text{if } m < x < 0.5 \\ \text{ambivalent} & \text{if } x = 0.5 \\ \text{tendentially-true} & \text{if } 0.5 < x < 1 - m \\ \text{true} & \text{if } 1 - m \leq x \end{cases} \quad \text{with } 0 \leq m < 0.5 \quad (3.60)$$

Categories 'true' and 'false' most closely represent the corresponding truth values of binary logic. Category 'tendentially-true' indicates that a FTV is in the upper half of the fuzzy value scale, but not close enough to represent a clear 'true.' The converse is the case for category 'tendentially-false.' Category 'ambivalent' covers the special case that a FTV is exactly between the extremes of 0 and 1, and thus there is no tendency towards either extreme.

We call m the *definiteness margin*. It determines the threshold between valuations categorised as 'true' or 'tendentially-true,' and 'false' or 'tendentially-false.' The value for m should to be defined by the process modeler. Since the process modeler also determines the valuators for FPVs about the tailoring context, with his choice of m he can provide a point of reference for the interpretation of FPV valuations: m determines the width of tolerance intervals at either end of the fuzzy value scale. Propositions with valuations within these tolerance intervals are to be considered true although their valuation is not quite 1, or false although their valuation is not quite 0, respectively.

With the five categories from (3.60) we can now adapt function *relevant* to accommodate fuzzy valuations. Prior to comparing valuations, we first map them to their respective categories:

$$relevant(\psi \wedge \omega) = \begin{cases} relevant(\psi) & \text{if } category(v(\omega)) \neq category(v(\phi)) \\ relevant(\omega) & \text{if } category(v(\psi)) \neq category(v(\phi)) \\ relevant(\psi) \cup relevant(\omega) & \text{otherwise} \end{cases}$$

$$(3.61a)$$

$$relevant(\psi \vee \omega) = relevant(\psi \wedge \omega) \qquad (3.61b)$$

$$relevant(\alpha) = \begin{cases} \{\alpha\} & \text{if } category(v(\alpha)) = category(v(\phi)) \\ \emptyset & \text{otherwise} \end{cases} \qquad (3.61c)$$

$$relevant(\neg\alpha) = \begin{cases} \{\alpha\} & \text{if } category(v(\neg\alpha)) = category(v(\phi)) \\ \emptyset & \text{otherwise} \end{cases} \qquad (3.61d)$$

To incorporate the previous definition (3.59) of function *relevant* for binary valuations, we extend the domain of *category* by binary truth values:

$$category(x \in \mathbb{B}) = x \tag{3.62}$$

Example 3.16 (relevance based on valuation categories) *Assume $\phi = \alpha \vee \beta$, and $v(\alpha)$ and $v(\beta)$ as in Example 3.15. With $m = 0.7$ and function 'relevant' as defined in (3.61), we have $category(v(\alpha)) = category(v(\beta)) = \text{true}$, so we get $relevant(\alpha \vee \beta) = relevant(\alpha) \cup relevant(\beta) = \{\alpha, \beta\}$.*

Justifications in Fuzzy Interval Logic

If we use *fuzzy interval logic* (FIL) instead of fuzzy logic, we can determine relevant atoms for valuations of FFs by simply providing a definition of function *category* for FTIs:

$$category([l, h] \in \hat{\mathbf{Q}}) = \begin{cases} \text{false} & \text{if } h < m \\ \text{tendentially-false} & \text{if } m \leq h < 0.5 \\ \text{ambivalent} & \text{if } l \leq 0.5 \leq h \qquad \text{with } 0 \leq m < 0.5 \\ \text{tendentially-true} & \text{if } 0.5 < l < 1 - m \\ \text{true} & \text{if } 1 - m \leq l \end{cases} \tag{3.63a}$$

and

$$\text{false} < \text{tendentially-false} < \text{ambivalent} < \text{tendentially-true} < \text{true} \tag{3.63b}$$

The range of the five categories is visualised in Figure 3.2. Any FTI in $\hat{\mathbf{Q}}$ maps to exactly one of the five categories. It may be completely in the range below m *(false)*, or completely in the range above $1 - m$ *(true)*. If not, it may at least be restricted to values below 0.5 *(tendentially false)*, or above 0.5 *(tendentially true)*. All other FTIs will include the middle value 0.5 and are thus fall in the category *ambivalent*.

Constructing a Justification Hierarchy

With our notion of relevance as defined in (3.61) and valuations in FIL we can now specify a justification hierarchy in **J** for every logic proposition ϕ with FTI valuation $\hat{v}$:

$$j_\phi = (\phi, \hat{v}(\phi), \{(\alpha, \hat{v}(\alpha), \emptyset) \mid \alpha \in relevant(\phi)\}) \in \mathbf{J} \tag{3.64}$$

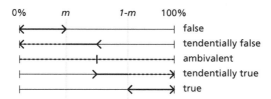

Figure 3.2: Segmentation of fuzzy intervals from $\hat{Q}$ into five intervals. Outward pointing arrowheads indicate inclusive interval bounds, inward pointing arrowheads indicate exclusive bounds. A FTI belongs to a class if it is fully contained in the respective interval and at least partly covers the region of the solid line. All FTIs that include FTV 0.5 fall within class "ambivalent."

Example 3.17 (justifying the valuation of a logic proposition) *With* $\phi = \alpha \wedge (\neg\beta \vee \gamma)$ *and* $\hat{v}(\alpha) = [0.6, 1]$, $\hat{v}(\beta) = [0.2, 0.4]$, $\hat{v}(\gamma) = [0.2, 0.7]$, *and* $m = 0.7$ *we get* $\hat{v}(\phi) = [0.6, 0.8]$ *and* $relevant(\phi) = \{\alpha, \beta\}$ *because* $category(\phi) = category(\alpha) = category(\neg\beta) =$ tendentially-true. *With* (3.64) *we get justification hierarchy*

$$(\phi, [0.6, 0.8], \{(\alpha, [0.6, 1], \emptyset), (\beta, [0.2, 0.4], \emptyset)\}) \tag{3.65}$$

3.2.3 Justifying Decision Ratings

In Section 3.2.2 we have developed a scheme to provide justifications for logic propositions valuated in FIL: Given a FF ϕ and a FTI valuation function $\hat{v}$, we can determine all atoms in ϕ which, along with their valuations, constitute the relevant facts that justify the valuation of ϕ.

The rating mechanism for tailoring decisions outlined in Section 2.4 relies on hypotheses about tailoring options. If these hypotheses are expressed as FFs (Section 2.4.2) and valuated in FIL (Section 2.4.5), we can provide justifications for the valuations of the necessitating and qualifying hypotheses of every tailoring option.

However, a process tailorer is not primarily interested in isolated hypotheses. He wants to understand why the TSS has assigned a specific rating to a tailoring decision. In Section 2.4.3 we have defined function $rating_C$ (2.40) to rate tailoring decisions. The rating is calculated in a different way depending on whether the tailoring option in question has been chosen, or excluded from the tailoring configuration. Therefore we must also distinguish these two cases when we justify a decision rating.

Justifying the Rating of a Chosen Tailoring Option

Let us first consider the case that a tailoring option has been chosen. From (2.40) and with $C(o) = \text{true}$ we get

$$rating_C(o) = \min(v_n(o), v_q(o) + k) \tag{3.66}$$

where k is the feasibility offset.

With hypotheses valuated as FTIs, we can apply function $rating_C$ to both the lower and upper bounds of the valuation intervals to obtain pessimistic and optimistic bounds for the decision rating (see functions $rating_C^{\text{pess}}$ and $rating_C^{\text{opt}}$ defined in (2.46) on page 33):

$$\widehat{rating}_C(o) = \left[rating_C^{\text{pess}}(o), rating_C^{\text{opt}}(o) \right] \tag{3.67}$$

As with justifying the valuations of logic propositions, we now again need to give an account of how an overall rating—in this case, the rating a tailoring decision—can be justified on the grounds of the valuations of its parts, i.e., the ratings of the necessitating and qualifying hypotheses.

In accordance with Section 3.2.2, the following assertions justify the valuations $\hat{v}_n(o)$ and $\hat{v}_q(o)$ of the necessitating and qualifying hypotheses for o:

$$j_n = (\text{``option } \langle o \rangle \text{ is necessary''}, \hat{v}_n(o), \{(\alpha, \hat{v}(\alpha), \emptyset) \mid \alpha \in relevant(hyp_n(o))\}) \tag{3.68a}$$

$$j_q = (\text{``option } \langle o \rangle \text{ is feasible''}, \hat{v}_q(o), \{(\alpha, \hat{v}(\alpha), \emptyset) \mid \alpha \in relevant(hyp_q(o))\}) \tag{3.68b}$$

In (3.68) and in all justifications presented in the following, expressions within angles "$\langle \ldots \rangle$" are to be rendered as appropriate textual representations, e.g., the name of option o for "$\langle o \rangle$."

We justify the application of feasibility offset k as follows:

$$j_{q,k} = \begin{cases} (\text{``} \langle o \rangle \text{ is feasible with offset } \langle k \rangle \text{''}, \hat{v}_{q,k}(o), \{j_q\}) & \text{if } k > 0 \\ j_q & \text{otherwise} \end{cases} \tag{3.69a}$$

where

$$\hat{v}_{q,k}(o) = \left[\min(lo(\hat{v}_q(o)) + k, 1), \min(hi(\hat{v}_q(o)) + k, 1) \right] \tag{3.69b}$$

To determine the impact of the necessity rating $\hat{v}_n(o)$ and the feasibility rating $\hat{v}_{q,k}(o)$ on the overall decision rating, we again check whether the constituting ratings are in the same category as the overall rating:

$$j_{C(o)} = \left(\text{``option } \langle o \rangle \text{ should be chosen''}, \widehat{rating}_C(o), J \right) \tag{3.70a}$$

where

$$J = \begin{cases} \{j_n\} & \text{if } category(\widehat{rating}_C(o)) \neq category(\hat{v}_{q,k}(o)) \\ \{j_{q,k}\} & \text{if } category(\widehat{rating}_C(o)) \neq category(\hat{v}_n(o)) \\ \{j_n, j_{q,k}\} & \text{otherwise} \end{cases} \quad (3.70\text{b})$$

Example 3.18 (justifying the rating of a chosen tailoring option) *Assume that for tailoring option o we have the hypotheses*

$$hyp_q(o) = \text{big-team} \quad (3.71\text{a})$$
$$hyp_n(o) = \text{big-team} \quad (3.71\text{b})$$

Thus, both the necessity and the feasibility of option o depend on the truth of 'big-team.' Let us further assume $\hat{v}(\text{big-team}) = [0.2, 0.4]$, a feasibility offset of $k = 0$, and a definiteness margin of $m = 0.3$. Assuming that o is chosen, with (3.66) and (3.67) we get $rating_C(o) = [0.2, 0.4]$ because its necessity and feasibility are limited by the valuation of big-team. Therefore we get justification

$$\begin{aligned} j_{C(o)} = (&\text{``option } \langle o \rangle \text{ should be chosen''}, [0.2, 0.4], \{ \\ &(\text{``option } \langle o \rangle \text{ is necessary''}, [0.2, 0.4], \{ \\ &(\text{``}\langle \text{big-team} \rangle \text{''}, [0.2, 0.4], \emptyset)\}), \\ &(\text{``option } \langle o \rangle \text{ is feasible''}, [0.2, 0.4], \{ \\ &(\text{``}\langle \text{big-team} \rangle \text{''}, [0.2, 0.4], \emptyset)\})\}) \end{aligned} \quad (3.72)$$

stating that it is rather undesirable that the option should be chosen (valuation $[0.2, 0.4]$), because the option is rather unnecessary and unfeasible (again by valuation $[0.2, 0.4]$), because the team is rather not big (again by valuation $[0.2, 0.4]$).

Justifying the Rating of an Excluded Tailoring Option

If a tailoring option has been excluded from the tailoring configuration, its rating depends only on the valuation of its necessitating hypothesis: From (2.40) and with $C(o) = false$ we get

$$rating_C(o) = 1 - v_n(o) \quad (3.73)$$

and justification

$$j_{-C(o)} = \left(\text{``option } \langle o \rangle \text{ should be excluded''}, \widehat{rating}_C(o), \{j_n\}\right) \quad (3.74)$$

with j_n as defined in (3.68a).

While justifications as in (3.74) are sufficient—an unnecessary option need not be chosen—it is still useful for the process tailorer to know if it would be feasible to choose it anyway. Therefore we extend (3.74) as follows:

$$j'_{-C(o)} = \left(\text{"option } \langle o \rangle \text{ should be excluded"}, \widehat{rating}_C(o), J \right) \qquad (3.75a)$$

where

$$J = \begin{cases} \{j_n, j_{q,k}\} & \text{if } category(\hat{v}_q(o)) \leq category(\hat{v}_n(o)) \\ \{j_n\} & \text{otherwise} \end{cases} \qquad (3.75b)$$

Example 3.19 (justifying the rating of an excluded tailoring option) *Assume we have the same hypotheses for option o as in Example 3.18, and $\hat{v}(\text{big-team}) = [0.2, 0.4]$, $k = 0$, and $m = 0.3$. Now if o is not chosen, with (3.73) and (3.67) we get $rating_C(o) = [0.6, 0.8]$, i.e., the 1-complement or $\hat{v}_n(o)$. Since $hyp_q(o) = hyp_n(o)$ we have $category(\hat{v}_q(o)) = category(\hat{v}_n(o))$ and therefore provide a justification that mentions not only the necessity, but also the feasibility of option o:*

$$\begin{aligned} j_{-C(o)} = (\text{"option } \langle o \rangle \text{ should be excluded"}, [0.6, 0.8], \{ \\ (\text{"option } \langle o \rangle \text{ is necessary"}, [0.2, 0.4], \{ \\ (\text{"}\langle \text{big-team} \rangle \text{"}, [0.2, 0.4], \emptyset)\})\}) \\ (\text{"option } \langle o \rangle \text{ is feasible"}, [0.2, 0.4], \{ \\ (\text{"}\langle \text{big-team} \rangle \text{"}, [0.2, 0.4], \emptyset)\})\}) \end{aligned} \qquad (3.76)$$

stating that it is rather desirable that the option should be excluded (valuation $[0.6, 0.8]$), because the option is rather not necessary (valuation $[0.2, 0.4]$), because the team is rather not big (valuation $[0.2, 0.4]$), and that the option is rather not feasible for the same reason.

3.2.4 Rating and Justifying the Quality of the Tailoring Context

With the introduction of general tailoring rules we have also broadened the notion of tailoring hypotheses: Every general tailoring rule that includes FPVs $\alpha_i = proposition(\dots)$ about the tailoring context (see (2.54) and (2.57) on page 36) also entails necessitating or qualifying hypotheses about each α_i.

Example 3.20 (hypotheses about the tailoring context) *Consider a tailoring rule stating that option o can only be configured when the team is big:*

$$O(configured(o) \rightarrow big\text{-}team) \qquad (3.77)$$

with FPV big-team defined as in (2.56) in Example 2.19. From this, we get the hypotheses

$$hyp_q(o) = \textit{big-team} \tag{3.78a}$$

$$hyp_n(\textit{big-team}) = \textit{configured}(o) \tag{3.78b}$$

and, assuming there are no other rules,

$$hyp_q(\textit{big-team}) = \text{true} \tag{3.78c}$$

$$hyp_n(o) = \text{false} \tag{3.78d}$$

With the aid of these hypotheses, we can determine whether the actual valuation of any FPV about the tailoring context is adequate, that is, whether it is in agreement with other aspects of the tailoring context, and with the current tailoring configuration. However, we cannot directly apply rating function $rating_C$ defined in (2.40): $rating_C$ is defined for only two cases—either an option is chosen, or not. When rating a FPV about the tailoring context, we need to base this rating on the fuzzy valuation of the FPV instead of the two discrete cases distinguished by (2.40). Therefore we first need to adapt our rating scheme to valuations of FPVs.

Rating the Tailoring Context

To rate valuations of FPVs about the tailoring context, we define function $rating_C$ for the extended range of FPVs in $\mathbf{A}$ (see (3.1) on page 53) as

$$rating_C : \mathbf{A} \mapsto \mathbf{Q} \tag{3.79a}$$

with

$$rating_C(\alpha) = 1 - penalty(v(\alpha), v(hyp_n(\alpha)), v(hyp_q(\alpha))) \tag{3.79b}$$

and

$$penalty(v_\alpha, v_n, v_q) = \max(0, v_n - v_\alpha, v_\alpha - v_q) \tag{3.79c}$$

Definition (3.79) is based on the following rationale: The degree to which a FPV $\alpha \in \mathbf{A}$ applies should not be lower than the degree of its necessity, i.e., it should be the case that $v(\alpha) \geq v(hyp_n(\alpha))$. Otherwise, the amount to which $v(\alpha)$ falls short of $v(hyp_n(\alpha))$ should be subtracted from the optimal rating 1 as a penalty. Similarly, the degree to which a FPV $\alpha \in \mathbf{A}$ applies should not be higher than the degree of its feasibility. Hence, it should be the case that $v(\alpha) \leq v(hyp_q(\alpha))$. Otherwise, the amount to which $v(\alpha)$ exceeds of $v(hyp_q(\alpha))$ should be subtracted from the optimal rating 1 as a penalty. If there are penalties for both necessity and feasibility, the higher penalty should come to play, therefore (3.79) calculates the maximum of both penalties. We also make sure in (3.79) that there are no negative penalties, since the optimal rating must not exceed 1.

Now the task remains to define ratings about the tailoring context in FIL based on valuations

$$\hat{v}_\alpha = [l_\alpha, h_\alpha] = \hat{v}(\alpha) \tag{3.80a}$$

$$\hat{v}_n = [l_n, h_n] = \hat{v}(hyp_n(\alpha)) \tag{3.80b}$$

$$\hat{v}_q = [l_q, h_q] = \hat{v}(hyp_q(\alpha)) \tag{3.80c}$$

Function $\widehat{rating}_C$ can be defined implicitly as follows:

$$\widehat{rating}_C : \mathbf{A} \mapsto \hat{\mathbf{Q}} \tag{3.81a}$$

$$\widehat{rating}_C(\alpha) = \left[1 - \max_{v_\alpha, v_n, v_q} penalty(v_\alpha, v_n, v_q), 1 - \min_{v_\alpha, v_n, v_q} penalty(v_\alpha, v_n, v_q) \right] \tag{3.81b}$$

where $l_\alpha \le v_\alpha \le h_\alpha, l_n \le v_n \le h_n, l_q \le v_q \le h_q$.

Because of the non-linear nature of function *penalty*, the implicit definition (3.81b) represents a non-linear optimisation problem with three variables. A formal derivation of an explicit definition for $\widehat{rating}_C(\alpha)$ is mathematically complex, but by taking into account a few properties of function *penalty*, we present a plausible solution to this problem in a more condensed form.

Our goal is to find the highest and the lowest penalty for any triple of fuzzy valuations (v_α, v_n, v_q) ranging within the interval valuations $\hat{v}_\alpha \times \hat{v}_n \times \hat{v}_q$. Let us first consider which triple maximises the penalty.

(3.79c) is based on a necessity penalty $v_n - v_\alpha$ and a feasibility penalty $v_\alpha - v_q$. To maximise the necessity penalty, we assume the highest possible necessity ($v_n = h_n$), and the lowest possible valuation for FPV α ($v_\alpha = l_\alpha$). Similarly, to maximise the feasibility penalty, we assume the lowest possible feasibility ($v_q = l_q$), and the highest possible valuation for FPV α ($v_\alpha = h_\alpha$). By this we have constructed the two "worst cases": Either α is the case to the greatest degree, but is also unfeasible to the greatest degree, or α is *not* the case to the greatest degree, but is also necessary to the greatest degree. Therefore, the highest achievable penalty is

$$\max_{v_\alpha, v_n, v_q} penalty(v_\alpha, v_n, v_q) = \max(0, h_n - l_\alpha, h_\alpha - l_q) \tag{3.82a}$$

Similarly, we can derive the lowest achievable penalty as

$$\min_{v_\alpha, v_n, v_q} penalty(v_\alpha, v_n, v_q) = \max(0, l_n - h_\alpha, l_\alpha - h_q) \tag{3.82b}$$

and thus get the following explicit definition for $\widehat{rating}_C$:

$$\widehat{rating}_C(\alpha) = [1 - \max(0, l_n - h_\alpha, l_\alpha - h_q), 1 - \max(0, h_n - l_\alpha, h_\alpha - l_q)] \qquad (3.83)$$

Example 3.21 (rating a FPV about the tailoring context) *Suppose that in a given tailoring context, the team size is estimated to be in the range between 6 and 7, so with valuator (2.55) we get* $\hat{v}(big\text{-}team) = [0.2, 0.4]$: *by definition of 'big-team,' the team is not really big. Suppose also that option o from Example 3.20 is chosen. With the hypotheses from Example 3.20 we get*

$$\hat{v}(big\text{-}team) = [0.2, 0.4] \qquad (3.84\text{a})$$

$$\hat{v}(hyp_n(big\text{-}team)) = \hat{v}(configured(o)) = [1, 1] \qquad (3.84\text{b})$$

$$\hat{v}(hyp_q(big\text{-}team)) = \hat{v}(\text{true}) = [1, 1] \qquad (3.84\text{c})$$

$$\widehat{rating}_C(big\text{-}team) = [1 - \max(0, 1 - 0.2, 0.4 - 1), 1 - \max(0, 1 - 0.4, 0.2 - 1)]$$
$$= [0.2, 0.4] \qquad (3.84\text{d})$$

So we conclude that given the current tailoring configuration, it is rather undesirable that the team size is rather not big. As is to be expected, this is reflected by a bad rating for the tailoring decision about o: We get $\widehat{rating}_C(o) = [0, 0]$ *because option o is configured, but by (3.78d) is not necessary.*

Justifying Ratings about the Tailoring Context

Now that we can rate the valuations of propositions about the tailoring context, we can proceed to construct the justification. As in Section 3.2.3, at the root of the justification hierarchy we have an assertion consisting of a statement and a rating for the statement, and accompanying the assertion we have justifications for the rating. With proposition α, we have

$$j_\alpha = \left(\text{"It is adequate that it is } \langle category(\hat{v}(\alpha)) \rangle \text{ that } \langle \alpha \rangle \text{.", } \widehat{rating}_C(\alpha), J \right) \qquad (3.85\text{a})$$

where J are the subordinate justifications for j_α.

We cannot define J in the same way as when justifying tailoring decision ratings: Because of the penalty concept introduced above, ratings about the tailoring context represent a relative distance between valuations for FPVs and their associated hypotheses. Therefore, comparing the truth categories of the rating with those of the hypotheses does not produce meaningful results. To determine whether the necessitating hypothesis for α, the qualifying hypothesis for α, or both have contributed to the rating for α, we calculate two additional auxiliary ratings for α that each consider only one of the two hypotheses:

$$\widetilde{rating}_C^n(\alpha) = [1 - \max(0, h_n - l_\alpha), 1 - \max(0, l_n - h_\alpha)] \tag{3.86a}$$

$$\widetilde{rating}_C^q(\alpha) = [1 - \max(0, h_\alpha - l_q), 1 - \max(0, l_\alpha - h_q)] \tag{3.86b}$$

If one or both of these auxiliary ratings is in the same truth category as $\widetilde{rating}_C(\alpha)$, we consider the associated hypothesis a contributor to the rating of α and include it in the justification hierarchy.

$$J = \begin{cases} \{j_\alpha, j_n\} & \text{if } category(\widetilde{rating}_C(\alpha)) \neq category(\widetilde{rating}_C^q(\alpha)) \\ \{j_\alpha, j_q\} & \text{if } category(\widetilde{rating}_C(\alpha)) \neq category(\widetilde{rating}_C^n(\alpha)) \\ \{j_\alpha, j_n, j_q\} & \text{otherwise} \end{cases} \tag{3.87a}$$

$$j_\alpha = (\text{``}\langle\alpha\rangle\text{''}, \hat{v}(\alpha), \emptyset) \tag{3.87b}$$

$$j_n = \left(\text{``it is necessary that } \langle\alpha\rangle\text{''}, \hat{v}_n(\alpha), \bigcup_{\beta \in relevant(hyp_n(\alpha))}\{(\beta, \hat{v}(\beta), \emptyset)\}\right) \tag{3.87c}$$

$$j_q = \left(\text{``it is feasible that } \langle\alpha\rangle\text{''}, \hat{v}_q(\alpha), \bigcup_{\beta \in relevant(hyp_q(\alpha))}\{(\beta, \hat{v}(\beta), \emptyset)\}\right) \tag{3.87d}$$

In addition to justifications about hypotheses, with j_α we also always include the actual valuation of FPV α as an additional justification for the rating of that justification.

Example 3.22 *In Example 3.21 we have calculated the adequacy of valuation $\hat{v}(big\text{-}team) = [0.2, 0.4]$. We have $\widetilde{rating}_C^n(\alpha) = [0.2, 0.4]$ and $\widetilde{rating}_C^q(\alpha) = [1, 1]$, so we conclude that only the necessitating hypothesis contributes to the rating for α. With (3.87) and definiteness margin $m = 0.3$ we get the justification hierarchy*

$$\begin{aligned} j_{big\text{-}team} = (&\text{``it is adequate that it is } \langle\text{tendentially-false}\rangle \text{ that } \langle big\text{-}team\rangle\text{''}, [0.2, 0.4], \{ \\ &(\text{``}\langle big\text{-}team\rangle\text{''}, [0.2, 0.4], \{ \\ &(\text{``it is necessary that } \langle big\text{-}team\rangle\text{''}, [1, 1], \{ \\ &(\text{``}\langle configured(o)\rangle\text{''}, [1, 1], \emptyset)\}\}\} \end{aligned} \tag{3.88}$$

$j_{big\text{-}team}$ justifies the low desirability $[0.2, 0.4]$ of valuation $category(\hat{v}(big\text{-}team)) = $ tendentially-false with the high necessity of that valuation. This in turn is justified by the fact that o is chosen in the tailoring configuration.

Justifications are only useful if they can be understood intuitively. So far we have developed justifications on a formal level. In the following section we will show how justifications such as in Example 3.22 can be presented to users of the TSS in a comprehensible fashion.

EXCLUDED: "use cases"
• option "use cases" should be excluded $[0.6, 0.8]$
• option "use cases" is necessary $[0.2, 0.4]$
• *big-team* $[0.2, 0.4]$
• option "use cases" is feasible $[0.2, 0.4]$
• *big-team* $[0.2, 0.4]$

Table 3.1: A simple representation of a justification hierarchy

3.2.5 Presenting Justifications in a GUI

The TSS constructs justifications for its suggested tailoring decisions in order to provide as much transparency as possible to the process tailorer. The form in which justification hierarchies are presented must therefore be as self-explanatory and easy to grasp as possible.

In the preceding sections, we have laid the foundations of such a representation by introducing a uniform hierarchic structure that serves to summarise the results of a complex, multi-layered rating mechanism as outlined in the introduction to Section 3.2.

We will now, in several steps, develop a graphical representation of justification hierarchies. To provide examples for graphical representations, we will use justification hierarchy (3.76) about excluded option o from Example 3.19:

$$j_{-C(o)} = (\text{"option } \langle o \rangle \text{ should be excluded"}, [0.6, 0.8], \{$$
$$(\text{"option } \langle o \rangle \text{ is necessary"}, [0.2, 0.4], \{$$
$$(\text{"} \langle big\text{-}team \rangle \text{"}, [0.2, 0.4], \emptyset)\})\})$$
$$(\text{"option } \langle o \rangle \text{ is feasible"}, [0.2, 0.4], \{$$
$$(\text{"} \langle big\text{-}team \rangle \text{"}, [0.2, 0.4], \emptyset)\})\})$$

As a first step, we can simply represent this justification hierarchy as a list of appropriately indented statements along with their grades of applicability. Assuming option o is named "use cases," Table 3.1 shows the representation for (3.76).

Table 3.2 shows a representation that is improved in several ways: First, we can add a clearer indication of the degree to which statements hold, and of whether an option is chosen or excluded. We therefore precede each statement with one of the symbols in Table 3.3 to indicate the truth category of the statement's rating as defined in (3.63). Additionally, statements that are marked 'false' or 'tendentially-false' are enclosed in square brackets to indicate that they do not hold. The numeric representation of the statement's applicability is replaced by a graphical range indicator where the applicability is marked as a black horizontal bar within a white region. On a colour display this black

○ EXCLUDED: "use cases"	
(✓) option "use cases" should be excluded	▭▮▭
(✗) [option "use cases" is necessary]	▭▮▭
(✗) [the team is big]	▭▮▭
(✗) [option "use cases" is feasible]	▭▮▭
(✗) [the team is big]	▭▮▭

Table 3.2: An extended representation of a justification hierarchy

symbol	state/truth category
○	excluded
●	chosen
✓	true
(✓)	tendentially-true
▲	ambivalent
(✗)	tendentially-false
✗	false

Table 3.3: Symbols for truth categories

bar can even be replaced by a clipping from a colour gradient that ranges from red at the far left to green at the far right. Lastly, if the process modeler is allowed to provide verbose counterparts for every FPV that refers to the context, the system can provide a clear statement instead of an internally used variable name such as *big-team* for assertions about the tailoring context.

With representations such as in Table 3.2, the process tailorer can quickly recognise which tailoring decisions are deprecated by the TSS by looking at the rating bars at the root level of the justification hierarchy for each tailoring decision. He should carefully check all tailoring decisions with rating symbols other than '✓'. To make this check easier, the application that displays a list of tailoring decisions should first hide the subordinate levels of all justification hierarchies and expand them only when the user wants to inspect the justification of a particular tailoring decision.

We conclude with an example for justification (3.88) about the tailoring context in Table 3.4. Note that enforcing a minimum width of the black bar ensures that also point intervals such as [1, 1] remain clearly visible.

Many more examples for representations of justifications can be found in Appendix C where we provide a completely justified tailoring configuration. We will discuss this in Chapter 4.3.

(✗)[it is adequate that it is tendentially false that the team is big]	▭
(✗) [the team is big]	▭
✓ [it is necessary that the team is big]	▭
✓ [option *o* is chosen]	▭

Table 3.4: Graphical representation of justification (3.88)

3.2.6 Outlook

In Section 3.2 we have laid the foundations for establishing and presenting justifications for ratings of both tailoring decisions and property valuations. Notwithstanding these considerations, the quality of the resulting justifications can still be improved.

At the heart of justifications are statements in natural language. They are associated with ratings, represented as bars, and are ordered hierarchically by a relation of the form "*x* because *y*." Generating expressive natural language statements is crucial for arriving at a comprehensible justification hierarchy. The solution presented so far has only limited possibilities with respect to that.

Justification (3.88) is a good example for this: The wording of the top level statement, "it is adequate that it is tendentially false that the team is big," is rather complicated. However, with the present means it is not possible to improve the wording, because the level at which the justification mechanism can combine statements is too coarse. For example, the FPV associated with this statement can only be represented by the unmodifiable phrase "the team is big." Better statements could be generated if FPVs were represented by more generic templates such as "the team is ⟨*mod*⟩ big" where the placeholder ⟨*mod*⟩ could be replaced by expressions "not," "rather not," "possibly," "rather," or by nothing, depending on the truth category for α. The top level statement of justification (3.88) might then read "it is adequate that the team is rather not big." We could then do without the square brackets used in our present approach to indicate a false statement. This consideration is only a simple example and a mere door-opener to natural language generation , which is in itself a broad field of research (see, for example, [RD00]).

Another area of improvement is the scope of ratings and justifications about the tailoring context. In Section 3.2.4, we have developed a method for rating *propositions* about the tailoring context that appear in the tailoring rules, and provide justifications only for these propositions. The scheme could be extended to directly rate and justify measurements of the tailoring context. This would provide yet more direct feedback on the quality of the tailoring context, but is not easy to realise since several different FPVs can refer to the same measurement.

4 Practical Applications of a Tailoring Support System

In Chapters 2 and 3 we have outlined all aspects of a TSS required to provide an implementation as a software tool. To demonstrate the ideas put forward in these chapters, we have implemented a TSS in the *Java* programming language [Sun]. To provide an experimental validation of our tailoring approach, we have constructed a sample case to demonstrate and verify its workings. This case is derived from the requirements engineering process framework currently under development by the *ReqMan* project consortium.

After giving a few comments on the TSS implementation in Section 4.1, we supply a brief overview of the *ReqMan* project in Section 4.2 before we discuss an extensive example case for process tailoring in Section 4.3.

4.1 Implementing the Tailoring Support System

In chapters 2 and 3 we have developed a tailoring framework that includes all necessary formal foundations of a TSS. The data structures, algorithms, and functions provided in these chapters are sufficient to permit an implementation of the tailoring framework as a software library.

We have chosen to implement the tailoring framework in the *Java* programming language. The implementation covers all aspects discussed in the preceding chapters: Given a tailoring universe and a tailoring context, it is able to produce a tailoring configuration with justified recommendations for each tailoring decision, and with ratings for the valuation of every proposition about the tailoring context that occurs somewhere in the tailoring rules. The library is also able to re-tailor partial configurations when some tailoring decisions have already been predetermined.

The data model of the library follows the formal definitions in Chapter 2. It includes an API[1] that allows other applications to access the functionality of the library. The library also provides an XML interface that makes it possible to provide tailoring universes, tailoring contexts, and partial tailoring configurations as files or streams in the XML

---

[1] application programming interface

format. The library can also write tailoring results to an XML file; an XSLT template is provided to transform the input to the tailoring algorithm as well as its output to HTML pages for easy review in a web browser. There is also an XSLT template that translates XML input and output to LaTeX code, allowing us to share some actual examples further down in Section 4.3, and, as research on our tailoring framework will continue, also in future publications.

The tailoring library has been developed with the goal in mind that it will be integrated with existing process management software, in particular with the web-based process portal project > kit [HMVV04, PK].

The tailoring library has already been evaluated with the formal definition of a process framework for requirements engineering developed in the context of the ongoing *ReqMan* research project. In Section 4.2, we will now take a closer look at *ReqMan*, and at the role the tailoring library plays in the context of this project.

4.2 The *ReqMan* Project

The objective of the *ReqMan* research project is to "provide methods and tools for quality- and reuse-oriented requirements management in *small and medium enterprises* (SMEs) by integrating aspects of project, quality, requirements and re-use management which have up to now been mostly examined separately" [RM, translated from German].

Among the projected results of *ReqMan* are a reference process framework for requirements engineering and management, geared especially towards the needs of SMEs, and software tools that support the application of that framework. The *ReqMan* project is still in progress, and is expected to conclude towards the end of 2006.

The TSS developed in the context of this thesis is going to be one of the principal tools of the *ReqMan* project. As an extension of the process portal project > kit, which is on the market since 2002, it allows for tailoring the *ReqMan* process framework to the specific requirements of a given project. In Section 4.3 we provide a detailed example of how the TSS can be applied to the *ReqMan* reference process. Before we proceed with the example, we give a short overview of the structure of the *ReqMan* process framework.

The *ReqMan* process framework is divided into five phases (Figure 4.1). It partitions requirements engineering activities into four consecutive phases: Requirements elicitation, requirements analysis, requirements specification, and requirements verification and validation. All superordinate managerial activities are collected in a fifth phase for requirements management that accompanies the whole life-cycle of requirements engineering.

Every phase comprises a set of *practices* that represent generic activities specific to that phase. For every practice, the *ReqMan* framework proposes one ore more *techniques* that implement the practice by means of a specific method. Some techniques may even

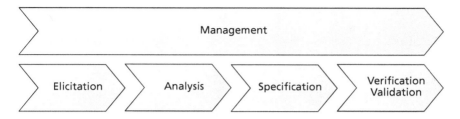

Figure 4.1: Phases of the *ReqMan* framework for requirements engineering processes

Figure 4.2: UML representation of the data model of the *ReqMan* framework

implement more than one practice at a time. Figure 4.2 summarises the basic data model of the *ReqMan* framework in a UML diagram.

Every practice is assigned one of three categories to indicate the relevance of the practice. [ODKE05] defines these categories as follows:

Basic represents a practice that is relevant in any requirements context.

Advanced represents a practice that is relevant in any requirements context, but requires other, usually basic, practices to be established.

Context represents a practice that is only relevant in certain project contexts.

As an example, Table 4.1 lists the practices for the elicitation phase. Figure 4.3 illustrates how techniques can be associated with these practices.

Elicitation Practice	Classification
Integrate stakeholders	base
Identify stakeholders and sources	base
Elicit functional requirements	base
Elicit goals	advanced
Determine scope	advanced
Elicit non-functional requirements	advanced
Elicit tasks and business processes	context dependent

Table 4.1: *ReqMan* practices for the elicitation phase

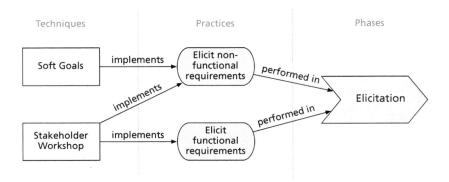

Figure 4.3: Techniques implementing practices from the *ReqMan* framework (illustration adapted from [ODKE05])

4.3 Example Tailoring of *ReqMan* Practices and Techniques

We now present a sample tailoring scenario to give an impression of how the TSS behaves in a practical setting. Appendix B details tailoring guidelines for a subset of the *ReqMan* process framework, which we will discuss in Section 4.3.1. Appendix C details a tailoring context and a justified tailoring recommendation calculated by the TSS on the grounds of that context, followed by ratings for the valuations of all propositions about the tailoring context that appear in the tailoring rules. We discuss the tailoring context in Section 4.3.2, and the results calculated by the TSS in Section 4.3.3.

4.3.1 The Tailoring Guidelines

Appendix B specifies tailoring guidelines for a subset of the *ReqMan* practices and techniques. We have restricted the example to practices for requirements elicitation. The example comprises definitions for entity types, tailoring options, and tailoring rules, and thus constitutes the input that would typically be provided by a process modeler. We now briefly go into each of these three kinds of definitions.

Tailoring Universe

Section B.1 specifies the entity types and their respective properties that constitute the characterisations of actual tailoring contexts.

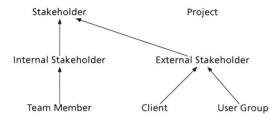

Figure 4.4: Hierarchy of entity types introduced in Appendix B

The principal types are the project itself, the members of the project team, and the client. Since our example is about picking the most appropriate techniques for requirements elicitation, we also introduce the "user group" as an additional entity type since the client commissioning new software is not necessarily identical with the user of the software, for example if the software is to be integrated into a product the client is selling.

Section B.1 also specifies some additional types marked as *abstract*. Also, some types specify that they *inherit* from another type. In short, we have a hierarchy of entity types. This is a feature of the TSS implementation that goes beyond the concept of entity types in Section 2.2.2, but which can be reduced to the original concept. To retain compatibility with the tailoring framework outlined in Chapter 2, we simply have to ignore all types marked *abstract*—these cannot be instantiated—and assume that all types t that inherit directly or indirectly from another type s also adopt all of the properties of s:

$$properties(t \in T) \supseteq \bigcup_{\{s|t \sqsubset^* s\}} properties(s) \qquad (4.1)$$

where $\sqsubset^*$ represents the transitive closure of the inheritance relation. Figure 4.4 provides an overview of the type hierarchy introduced in Appendix B.

Only few properties are defined for the entity types. Projects, for example, are characterised by their duration, the size of the team, the expected number of features, the criticality of the application domain, the product type (GUI, embedded, or COTS[2]), and the availability of relevant documentation. These properties are fairly general within the scope of software projects, maybe with the exception of the "relevant documentation" property which is more specifically coined towards the domain of requirements engineering. An additional property is of a more technical nature and specific to the structure of the *ReqMan* framework: It determines whether only basic practices should be selected, or whether advanced practices should be included. Stakeholders have even fewer properties, such as the team's or the client's experience and innovativeness, such that the total number of properties across all entity types amounts to only 17.

---

[2]commercial off-the-shelf [Lau02]

$$
\begin{array}{ll}
\text{``Ramp: } r\text{--}s\text{''}^3 \quad g(x) = ramp_{r,s}(x) = \begin{cases} 0 & \text{if } x < r \\ \frac{x-r}{s-r} & \text{if } r \leq x < s \\ 1 & \text{if } x \geq s \end{cases} \\[2em]
\text{``At least: } r\text{''} \qquad\qquad g(x) = \begin{cases} 1 & \text{if } x \geq r \\ 0 & \text{otherwise} \end{cases} \\[2em]
\text{``\textit{At most: } r''} \qquad\qquad g(x) = \begin{cases} 1 & \text{if } x \leq r \\ 0 & \text{otherwise} \end{cases} \\[2em]
\text{``\textit{Equals: } r''} \qquad\qquad g(x) = \begin{cases} 1 & \text{if } x = r \\ 0 & \text{otherwise} \end{cases}
\end{array}
$$

Table 4.2: Valuation functions for variable definitions in Section B.1

Each property defined in Section B.1 is accompanied by one or more definitions of FPVs. Each FPV specifies a valuation function that transforms the value of its associated property to a FTV. Table 4.2 shows the four valuation functions applied in Section B.1.

We have compiled the particular set of properties for *ReqMan* such as to obtain the best balance between two opposing forces: On the one hand, the relevant aspects of any entity should be documented as completely as possible. On the other hand, many properties will be relevant for only few tailoring decisions, so for every improvement of the quality of tailoring recommendations, an increasing number of properties must be specified. With too many properties, the complexity of making tailoring decisions will not be eliminated, but merely shifted to handling an equally complex tailoring context.

No real-world situation can be completely mapped to standardised categories. Since the TSS should not substitute the process tailorer but instead should support him in the best possible way, the tailoring universe should not be optimised for detailedness, but for usefulness, the latter of which we obtain by balancing complexity and coverage of entity properties.

Other published tailoring guidelines based on context properties use similar core properties, such as team size and application criticality [BR88, IEC98, Jal99, GQ94, Coco2, BT03], project duration [GQ94], or team skills [Jal99, BT03].

Tailoring Options

Section B.2 defines three kinds of tailoring options. First, it defines all practices for requirements elicitation from the *ReqMan* repository as reproduced in Table 4.1 on page 91. Second, there are various techniques for realising the practices. Since at present *ReqMan*

---

[3]For a detailed discussion of the ramp function, refer back to Section 2.4.1 on page 24.

does not yet provide a complete and documented collection of techniques, we have collected suitable techniques from other sources, mainly from [Lau02]. Finally, the last few tailoring options play a special role. They do not represent process elements, but hypotheses about the process that must be fulfilled if the corresponding option is chosen. We have not realised them as properties of the tailoring context because not only the process tailorer, but also the tailoring algorithm should be able to demand that a process hypothesis be fulfilled, on the grounds of the tailoring context: When process hypotheses are represented as tailoring options, the tailoring algorithm can recommend them if appropriate. We will discuss an example of a hypothesis in Section 4.3.3.

Tailoring Rules

The tailoring rules are realised internally as logic terms with standard operators as described in Section 2.3.2 on page 20. However, in order to make tailoring rules more legible and comprehensible, the rules in Section B.3 have been rendered in formalised English phrases representing those logic terms.

For example, a logical disjunction $A \vee B \vee C$ is rendered as

> **at least one** of:
>
> - A
> - B
> - C

and a negated disjunction as

> **none** of: [...]

Logic conjunctions are handled similarly.

Constructs of the form

> **if** A
>
> **then** it should be the case that B
>
> **else** it should be the case that C

have the logical equivalent $(A \rightarrow B) \wedge (\neg A \rightarrow C)$.

There are three kinds of atomic expressions in the tailoring rules: Logic constants, FPVs defined for entity types, and references to tailoring decisions.

Logic constants 'true' and 'false' are represented by the terms "always" and "never."

The application of FPVs v for property p of entity type t is rendered as

> **every** t: v

and represents the generic FPV *proposition*(t, p, g) as defined in (2.54) on page 36, where g represents the valuation function defined for v. Cardinality constraints as in *proposition*(t, p, g, c) (2.57) are also supported and are rendered as

> **an arbitrary** t: v

for $c = 1$, and as

> **an arbitrary** t **group of size** c: v

for $c > 1$.

Lastly, FPVs of the form *configured*(o) (cf. (2.35) on page 26) with tailoring option o are rendered as

> **chosen:** o

and negations of the form $\neg$*configured*(o) are rendered as

> **excluded:** o

We have now given an overview of how tailoring rules are represented in Section B.3. We will look at individual rules in Section 4.3.3 when we will discuss the tailoring recommendation calculated by the tailoring algorithm.

4.3.2 The Tailoring Context

Section C.1 represents a sample tailoring context that a process tailorer might have provided on the basis of the tailoring types and their properties in Section B.1. It defines several entities, each of which is represented in a separate table preceded by a heading specifying the type and name of the entity.

For example, the first entity, defined on page 155, is of type *Client* and has the name *BigCorp*. The definition table for each entity has three columns; each line represents the definition of a property whose name is given in the first column. The second column serves as a reminder of the scale required to measure a value for the property, and the third column states a measurement for the property, either as an exact value or as an interval. Exact values are implicitly transformed to point intervals by the TSS such that each measurement is available as an interval.

The TSS operates on FIL and depends on FTI valuations of interval measurements from the tailoring context. However, the valuation functions specified for the variable definitions map exact measurements to simple FTVs instead of FTIs. However, the TSS implicitly derives a FTI valuation function $\hat{g} : D \mapsto \hat{Q}$ from each simple valuation function $g : D \mapsto Q$ as outlined in Section 2.4.5 on page 31.

Example 4.1 (Transformation of a FPV valuation function) *Entity type* Project *in Section B.1 defines the following valuation function for variable "the duration is short":*

$$g(x) = ramp_{12,6}(x) \tag{4.2}$$

From this, and with (2.44), the TSS derives the FIL valuation function

$$\hat{g}([r, s]) = \begin{cases} [ramp_{12,6}(r), ramp_{12,6}(s)] & \textit{if } ramp_{12,6}(r) \leq ramp_{12,6}(s) \\ [ramp_{12,6}(s), ramp_{12,6}(r)] & \textit{otherwise} \end{cases} \tag{4.3}$$

Note that the tailoring context in Section C.1 does not define every available property for every entity. The TSS can deal with such incomplete information since function $\hat{g}$ valuates variables referring to unspecified properties to the FTI [0, 1] (cf. (2.44b) on page 31).

4.3.3 The Tailoring Result

Up to now, we have only discussed data that has been provided by users of the TSS, namely the process modeler and the process tailorer. With the tailoring guidelines and the tailoring context we have covered all necessary input required by the TSS to provide a tailoring recommendation, that is, a recommendation about which tailoring options should be chosen and which options should be excluded. Section C.2 reproduces the tailoring configuration recommended by the actual implementation of the TSS based on the input in sections B and C.1. An overview of all 41 tailoring decisions, along with their individual ratings, is followed by a detailed listing of justifications for the ratings of each decision. The *Java* implementation of the TSS took 7.8 seconds to calculate the tailoring configuration in Section C.2 on an Apple Mac Mini running at 1.4 GHz on a G4 processor. Only 171 out of the approximately 2.2 trillion (2^{41}) possible tailoring configurations had to be examined until the optimal configuration was found. The justifications are rendered as described in Section 3.2.5.

In the given example, all tailoring decisions, either for or against a tailoring option, have been put forward by the TSS except one: The hypothesis "Advanced elicitation practices should be carried out" has been manually included in the tailoring configuration before the tailoring algorithm was started. A comment preceding the justification for that hypothesis reveals that the hypothesis has been chosen by the user, and not by the TSS.

We now pick out some of the recommended tailoring decisions to highlight some typical interactions between the recommendations of the TSS, the underlying rules, and the tailoring context.

Practices

Rules 3–12 determine when a specific practice *must* be chosen (rules 3–5), and when it *can* be chosen (rules 6–12). Among the seven practices, three are base practices (Rule 3), three are advanced practices (Rule 4), and the remaining practice is context-dependent (Rule 5) and will only be recommended by the TSS if the project produces a GUI or COTS application.

Whether advanced practices are to be used is determined by rules 1 and 2. Rule 1 states that the advanced practices are necessary if the application domain is critical. Rule 2 states that the advanced practices are feasible only if some team member has basic experience with requirements management.

As noted above, in the given example the user has decided that advanced practices are to be included by choosing the corresponding tailoring option. The TSS respects this decision but assigns it a bad rating because it does not consider it necessary—the application domain is not critical. Nonetheless, as a consequence of the user's decision, the TSS not only recommends all three base practices, which are always required by Rule 3, but also recommends the three advanced practices. As can be seen from the justifications for recommended practices, the TSS has also ensured their feasibility by selecting suitable techniques realising these practices.

Hypotheses

As described in Section 4.2, the *ReqMan* project has established the two-level model of abstract practices and concrete techniques to better organise the techniques into categories and to ensure that no important aspect of requirements management is overlooked.

However, while this model proposes techniques to realise every practice, it does not provide clear criteria to determine whether a given selection of techniques provides sufficient coverage of the practices—in many cases, it will not be sufficient to choose just one technique for every practice. Tailorable hypotheses can help to ensure complete coverage by establishing minimal sets of complementary techniques, at least one of which is required for complete process coverage. For example, to ensure that requirements have been elicited completely, [Lau02] on page 338 recommends six techniques specifically suited to that purpose.

This recommendation is reflected in Rule 33 for hypothesis "Ensure completeness of elicitation," and Rule 34 states that the hypothesis should be recommended by default—it is always necessary to ensure complete requirements elicitation.

Double-checking uncertain decisions

Most tailoring options represent concrete elicitation techniques. In our case, all choices of techniques have been put forward by the TSS. When the process tailorer reviews the tailoring decisions recommended by the TSS, he should first verify all decisions whose rating does not fall within category 'true' (✓), i.e., where the corresponding recommendation statement supporting the tailoring decision is considered at most 'tendentially-true.'

In our example, four tailoring decisions demand special attention: The TSS considers it merely *tendentially true* that the techniques *Quality Function Deployment* and *Goal-Domain Analysis* should be excluded. A quick review of the justifications for these verdicts reveals that this doubt stems from the uncertainity about whether the duration of the project should be considered short. Also, the available formal competency might not be sufficient to allow for any of these two techniques. The remaining two tailoring decisions that should be reconsidered are hypotheses: The TSS advises against carrying out advanced elicitation practices for the reasons discussed above. Finally, the TSS does not fully endorse its recommendation to ensure that risks and consequences be elicited. Rule 32 states that the hypothesis holds if there are many features to be realised but the project's duration is short. While the TSS acknowledges that many features have to be realised, it is again the uncertainty about the duration of the project being short that reduces the overall weight of this recommendation.

Completing the Tailoring Configuration

The aim of the TSS is to find a tailoring configuration that best satisfies the tailoring rules. However, the process tailorer might have specific reasons to reject options recommended and chosen by the TSS, or to choose additional options. He can therefore modify the recommended tailoring configuration at will, and then repeat the automated tailoring process for all remaining options he has not decided about explicitly to find out how his decisions fit into the framework of tailoring rules. But even before changing any of the decisions proposed by the TSS, he can anticipate potential problems by looking at the justifications provided for the recommendations of the TSS.

In fact, the process modeler will in most cases not intend, nor be able, to cover every aspect that will need to be considered when tailoring a process, but will instead attempt to provide tailoring guidelines that address most of the common tailoring issues. Therefore, manual adaptations by the process tailorer will rather be the norm and not the exception.

In the given example, the TSS has excluded many options because it does not consider them feasible for various reasons. For example, it recommends against *Brainstorming* because of reportedly bad visionary skills, and it recommends against the *Task Demonstration* technique because it might not be possible to visit the users in their surroundings. The TSS has excluded other options only because they are not necessary: The TSS does not recommend the *Lauesen Interview* technique because it already recommends the *Lauesen Group Interview* technique, and one of the two is sufficient to cover the practice *Elicit Goals*.

4.3.4 Judging the Tailoring Context

The TSS provides a justified rating for the valuations of all propositions that occur in one or more tailoring rules. These ratings provide a different perspective on the quality of the overall tailoring situation: While ratings and justifications for tailoring decisions help the process tailorer evaluate what could be improved about the tailoring configuration, ratings and justifications about propositions help to identify problematic aspects of the tailoring context. In other words, the list of proposition ratings provides answers to the question of how the tailoring context should change to obtain a better rating for the current tailoring configuration.

In Section C.3, the TSS has declared two facts about the tailoring context more or less inadequate: First, the TSS considers it rather inadequate that not every team member has basic experience with requirements management. We already know from Section 4.3.3 that this is in conflict with the demand that advanced elicitation practices are to be used, and this observation is confirmed here by the justification for the associated proposition. Second, the TSS states that it is only tendentially adequate that many features will have to be realised in the course of the project. The justification reveals that the reason for this evaluation stems from the relative uncertainty about whether the duration of the project is short.

4.3.5 Outlook

In the above example, we have applied our tailoring approach to the *ReqMan* process framework, which has a particular structure involving specific kinds of dependencies between tailorable elements on three hierarchy levels, i.e., techniques, practices, and phases. This structure is not inherent in our tailoring framework itself, instead we used tailoring rules to express the dependencies inherent in the *ReqMan* process framework. The mechanism could also be used to tailor process models with altogether different structures, or, for that matter, any configuration problem outside the field of process models, as long as options can be identified that require binary decisions. For example, some process methodologies define tailoring as choosing values for *attributes* of process elements. We will treat these and other approaches to tailoring in Chapter 5, along with a discussion of how they could be integrated with our tailoring framework.

This chapter has only touched upon the possibilities a full-fledged graphical user interface to the TSS could offer. An appropriate interface would help the process tailorer to create and edit tailoring contexts. It could also present tailoring results in a much clearer way, for example by making use of colours and by allowing the user to vary the level of detail for justifications by providing handles to show or hide lower levels of the justification hierarchy. The system could also, on demand, display the qualifying and necessitating hypotheses which it has derived for every tailoring option, and for every proposition from the original tailoring rules. In addition, an appropriate user interface would encourage the interactive and iterative nature of the software-assisted tailoring process: The system provides an initial tailoring configuration which the user then changes in part before letting the TSS re-tailor the rest, repeating this cycle as often as needed.

The tailoring application could even perform its first tailoring before having received any information from the process tailorer. It could then judge from the grade of uncertainty of the property ratings what details of the tailoring context are especially relevant for its tailoring decisions. It would then first ask the process tailorer about those properties of the tailoring context that have received the widest, and therefore vaguest, rating intervals.

A process modeler would equally welcome tool support for maintaining extensive tailoring guidelines. This could include inspection tools to check the guidelines for gaps and contradictions. For example, one way to detect contradictions is to invoke the optimisation algorithm with a completely unspecified tailoring context. If tailoring decisions or context ratings still get optimistic ratings below 100%, then the process modeler can tell that the tailoring rules contain some insurmountable contradictions, since specifying data for the tailoring context will further restrict the range of possible ratings, but will never improve a decision rating. The process modeler will also need to structure and modularise large bodies of tailoring guidelines. For example, a family of similar process models might share a common subset of entity types and tailoring rules, while other entity types and rules are specific to particular extensions. An appropriate tool will have to provide the necessary infrastructure for such a modular approach.

Although many improvements are possible by providing an adequate application interface to the TSS, we have already illustrated the basic usefulness of the TSS: While the tailoring rules may become numerous and complex, the TSS still retains transparency and consistency by providing justifications of limited complexity for all of its recommendations and valuations. The ratings for tailoring decisions and proposition valuations complement each other and together help the process tailorer decide where and when he needs to improve his tailoring decisions, or the context of the project at hand, or both, and therefore provide valuable assistance in dealing with the complex task of process tailoring.

5 Related Approaches to Tailoring Assistance

So far we have presented our approach to a formal tailoring framework where a tailoring universe, tailoring contexts, and tailoring hypotheses provide the input for a search algorithm identifying the optimal tailoring configuration with regard to the given criteria. We have then illustrated how tailoring rules in deontic logic can be used to express tailoring guidelines from which tailoring hypotheses can be derived. To conclude, we have discussed an extensive example application of our tailoring approach.

Our results build on different fields of research, including techniques of *Artificial Intelligence* (AI) and various disciplines of logic. Our goal, however, was to contribute to the field of software process tailoring, and it is here that we were looking for new solutions. The methods of AI we applied and adapted to our purposes merely provided the tools, and we have provided references to the standard literature wherever we have introduced an AI concept. We therefore now focus our survey of related work on other approaches to process tailoring assistance, and will not delve further into similar applications of the AI techniques we have applied.

Our survey is divided into three parts. In Section 5.1 we examine the role of tailoring in established process reference models and process frameworks. In Sections 5.2 we shift our focus from established frameworks to tailoring approaches that set themselves apart by the level of formality at which they provide guidance for tailoring, and compare them with our own approach. Finally, in Section 5.3, we take a particularly detailed look at the German V-Model XT, which is both an established national software process standard, and at the same time takes tailoring by formal guidelines and even tailoring tool support further than any other presently established process framework.

We will draw some final conclusions from our survey in Section 5.4.

5.1 The Role of Tailoring in Process Reference Models and Process Frameworks

Among the most elaborate and universal approaches to formal software development processes relevant today are the process framework *Rational Unified Process* (RUP), and, on a much higher level of abstraction, the process reference and assessment models *Capability Maturity Model Integration* (CMMI) and *Software Process Improvement and Capability dEtermination* (SPiCE). Even though they differ greatly in their levels of abstraction and fields of application, the common goal of all three approaches is to address all aspects of software engineering as completely as possible. Consequently, they also have in common that they exhibit a degree of complexity that will in most circumstances require adjustments, mostly by reduction. Therefore, each of these approaches has developed some notion of process tailoring, of which we will now give a brief overview.

Another well-established process framework besides the RUP, albeit only in Germany, is the V-Model XT. Due to its particular relevance to the ideas of our approach we will deal with it separately in Section 5.3.

5.1.1 Capability Maturity Model Integration (CMMI)

CMMI [SEI02b] is a process reference model for software and systems engineering. It is the result of merging and integrating three previously independent *Capability Maturity Model* (CMM) reference models for systems engineering and software engineering. As a reference model, it does not provide actual process descriptions by itself, but instead provides a generic *model* for processes with the aim of providing a guideline for designing and assessing processes. To this end, CMMI specifies abstract characterisations of activities, called *practices* in the CMMI terminology, and groups related practices in *process areas*.

CMMI envisages tailoring on two levels of abstraction—tailoring actual descriptions of an organisation's standard processes, and tailoring the CMMI model itself. We now discuss both notions one at a time.

Process tailoring

CMMI defines processes as consisting of "activities that can be recognized as implementations of practices in a CMMI model" [SEI02b, p. 28]. Among various kinds of *organisational process assets*, CMMI comprises *tailoring guidelines*, which are intended to

> "[...] aid those who establish the defined processes for projects. Tailoring guidelines cover (1) selecting a standard process, (2) selecting an approved life-cycle model, and (3) tailoring the selected standard process and life-cycle model to fit project needs. Tailoring guidelines describe what can and cannot be modified and identify process components that are candidates for modification." [SEI02b, p. 26]

The basic act of process tailoring according to CMMI is as follows: An organisation has one or more *Organization's Standard Software Processes* (OSSPs), which define all relevant activities, and one or more *life-cycle models*, which group activities into sequential phases. Individual projects derive their *defined processes* from these standard processes and life-cycle models according to the organisation's tailoring guidelines. Selected standard processes are each tailored to corresponding defined processes. A defined process is a process that is actually enacted in the course of a project. A project may have more than one defined process; for example, one for development and one for testing. CMMI as a generic process reference model only goes as far as demanding that tailoring guidelines be supplied. As to be expected, it does not get specific about how tailoring should actually be carried out. However, approaches to establishing tailoring guidelines have been proposed in the context of CMMI [GQ94, Jal99]; we will discuss these in Section 5.2.

Model tailoring

CMMI defines *model tailoring* as "a process whereby only a subset of a model is used to suit the needs of a specific domain of application" [SEI02b, p. 87] with the intent "to assist an organization or project in aligning the CMMI products with its business needs and objectives, and thus focusing on those aspects of the products and services that are most beneficial to the organization" [SEI02b, p. 88]. In short, in the process of introducing CMMI or extending its adaptation of CMMI, an organisation may initially choose to focus only on a subset of the CMMI model, i.e., to *tailor* CMMI.

Model tailoring is relevant from two points of view—it can be applied to facilitate process improvements, i.e., constructive applications of the CMMI, or it can be applied to facilitate process benchmarking and appraisal, i.e., analytical applications of the CMMI. For each of these perspectives on tailoring, CMMI provides tailoring criteria of varying degrees of formality, ranging from general statements such as that "the exclusion of a significant number of process areas, goals, or practices may diminish the benefits achieved" [SEI02b, p. 89], to very specific rules such as that "goal" process elements must not be excluded if their respective process areas are included. The most immediate tailoring guideline consists in the categorisation of process areas into five *maturity levels*, where the inclusion of any particular process area requires the inclusion of all other process areas that are at a lower maturity level.

All explicit and implicit guidelines for model tailoring discussed above represent structural dependencies and do not refer to some kind of project or organisational context. This is justified considering that CMMI as a process reference model operates on an abstract level.

Given that the complete official documentation of CMMI occupies over 600 pages,[1] the four pages of advice on model tailoring it contains are comparatively slim. Consequently,

---

[1]counting only one of the two available representations, *staged* or *continuous*

a certain level of experience is necessary to perform model tailoring. While the authors of CMMI concede that model tailoring might be necessary especially for smaller projects, they maintain—in a slightly discouraging tone—that "model tailoring should only be done knowing that it can result in significant gaps in efforts to improve or appraise an organization's or a project's capabilities." [SEI02b, p. 88]

5.1.2 Software Process Improvement and Capability dEtermination (SPiCE)

SPiCE was published in 1998 as technical report ISO/IEC TR 15504 [ISO98a], and has been almost completely republished between 2003 and 2005 as an ISO/IEC standard [ISO05]. It provides a universal framework for improving and assessing processes by defining a two-dimensional *reference model* comprising a *process dimension* and a *capability dimension*.

The process dimension is derived from the companion standard ISO/IEC 12207, first published in 1995, and defines abstract processes related to software development. The framework is structured hierarchically into process categories such as *Engineering* and *Management*, which each comprise *first-level processes* which in turn specify subordinate *second-level processes*. Each process represents an activity that is characterised by descriptions of its purpose and outcomes.

The capability dimension defines six incremental levels of *process capability* and associates each level on this scale with *process attributes*. Process attributes are "features of a process that can be evaluated on a scale of achievement, providing a measure of the capability of the process" [ISO98b]. An example for a process attribute is the "process definition attribute," indicating the degree to which the process is defined with regard to a standard process. Process attributes are rated on a scale ranging from 0% *(not achieved)* to 100% *(fully achieved)*. Achieving a capability level is synonymous with achieving a rating above 50% for all process attributes associated with that level, and above 85% for all process attributes associated with a lower level.

As with CMMI, SPiCE includes the notion of *process tailoring* within its specification of the *process establishment process*, and demands that processes with a capability level of at least three "[include] appropriate guidance on tailoring" [ISO98b], albeit without further specifying what that guidance should look like. With regard to adapting the reference model itself, SPiCE provides no explicit correspondent to *model tailoring* as put forward by the CMMI. Still, SPiCE supplies several kinds of information that facilitate reducing and adapting the complexity of the full reference model to the requirements of a given project or organisation. These include the capability classifications for processes mentioned above, which effectively introduce dependencies between processes, and a comprehensive mapping between processes and associated work products [ISO98b, Annex A].

In practice, the lack of a *model tailoring* concept in SPiCE is reflected by the common practice of publishing variants of the ISO/IEC 15504 standard that are adapted to the requirements of particular industries, such as *Automotive SPiCE* developed by the German *Herstellerinitiative Software*[2] (HIS), a consortium of vehicle manufacturers [HIS], or *SPiCE for Space* [CVW+01], sponsored by the European Space Agency (ESA).

5.1.3 Rational Unified Process (RUP)

The RUP [IBM] is a commercial framework for iterative software development processes. The *Rational Software* company introduced it in 1995 as a more detailed variant of the *Unified Software Process* [JBR98]. The roots of the RUP date back to as far as 1987, when Ivar Jacobson created the first version of his *Objectory Process* [J+92] which focused on object-oriented development. IBM acquired Rational Software in late 2002. The RUP is supplied to its licensees in digital form along with a suite of software tools that support managing and enacting it. Due to this kind of distribution, the RUP is updated on a regular basis.

The RUP refers to process tailoring as "configuring the process." The RUP provides for configuring the software engineering process on two levels [Kru00, pp. 258]:

1. An organisation can have one or more *organisation-wide processes* that serve as templates for processes being enacted for actual projects. Process engineers can adapt these to the domains of application and core technologies used by their organisations. However, the RUP claims to be an appropriate organisation-wide process without further alterations.

2. The organisation-wide process can be tailored into a project-specific process, called a *development case* in RUP terminology. This kind of tailoring takes into account the context of an actual project.

The RUP is a very comprehensive framework that contains more than 300 elements [Hop03] organised into more than 100 selectable *process components*. Configuring the process thus is a complex task.

One of the tools accompanying the RUP distribution is called the *RUP Builder*. Its purpose is to support the act of configuring the process. It can be used to browse the process, select or de-select process components, and access a collection of pre-configured process templates. The RUP aims to facilitate its configuration by organising all of its process elements into a nested structure of process components, and also allows sets of components to be grouped together in a *process plug-in*. However, apart from providing structure to make navigating the process framework easier, the RUP Builder does not provide suggestions or guidelines about which process elements or components should be included in the

---

[2]Manufacturer's Software Initiative

process configuration, and it does not check whether the configured process is consistent and complete.

5.2 Formalised Tailoring Assistance

Formalised approaches to tailoring assistance distinguish themselves from other approaches in that their tailoring guidelines can be interpreted with little or no ambiguities, and that the application of these guidelines can be operationalised or even implemented as a software tool. We have observed two basic kinds of formalised tailoring assistance:

1. *rule-based approaches*, where tailoring recommendations are derived by applying a pre-defined set of static rules to an—at least partially—formalised description of the current project context to infer decisions about available tailoring options, and

2. *case-based approaches*, where a formalised description of the current project context is matched against exemplary contexts, or contexts of previous projects, stored in a database. Each context in the database is associated with a recommended tailoring configuration, and a query for a new context returns the configuration for the closest match.

We now discuss some realisations of both kinds of approaches in sections 5.2.1 and 5.2.2, and then compare their strengths and weaknesses in Section 5.2.3.

5.2.1 Tailoring by Inference: Rule-Based Approaches

Our own approach is a rule-based one—some given information is applied to pre-defined rules in order to derive recommendations for tailoring decisions. Another approach sharing this basic idea is the tailoring mechanism of the V-Model XT, which, for its special relationship with our own proposition, we will discuss separately and in more detail in Section 5.3.

IEC 61508

An easy way to represent tailoring rules is to map tailoring options to one or more categories by means of a two-dimensional matrix, or table. A prominent example for this kind of tailoring guidance is the safety norm IEC 61508 [IEC, IEC05] "Functional safety of electrical/electronic/programmable electronic safety-related systems." It specifies requirements for safety critical systems, recommending different techniques and measures—such as data flow diagrams, formal methods, or finite state machines—for the development of particular kinds of such systems. To decide which techniques and measures should be used under which circumstances, IEC 61508 distinguishes four *Safety Integrity*

Technique/Measure	Ref	SIL1	SIL2	SIL3	SIL4
1 Probabilistic testing	C.5.1	—	R	R	HR
2 Dynamic analysis and testing	B.6.5 Table B.2	R	HR	HR	HR
3 Data recording and analysis	C.5.2	HR	HR	HR	HR
⋮	⋮	⋮	⋮	⋮	⋮

Table 5.1: Excerpt from a SIL table [IEC98, Table A.5]

Levels (SILs) ranging from SIL1 (low) to SIL4 (very high). It specifies different situation-dependent formal methods for establishing the SIL of a current project. IEC 61508 provides tables [IEC98, Annex A] that recommend techniques and measures depending on the SIL of the project, as exemplified in Table 5.1. For each technique or measure and for each SIL, the recommendation is expressed as one of the four categories *highly recommended* ('HR'), *recommended* ('R'), *no recommendation for or against* ('—'), and *not recommended* ('NR').

IEC 61508 is not a complete process norm. It only addresses the safety dimension of processes. From this point of view, it is justifiable that there is only a single attribute—the SIL—to characterise the tailoring context. Also, the tailoring guidelines get by with a simple structure; every tailoring option can be considered independently of the others, based exclusively on the SIL. All this makes it easy to express tailoring guidelines in a simple matrix, mapping values of the SIL attribute to recommendations about available tailoring options. However, it also shows that the approach taken by IEC 61508 is very specific to its purpose and cannot be applied to more complex tailoring scenarios that require multiple context properties and more elaborate tailoring rules.

On the other hand, our tailoring framework is generic enough to emulate the SIL tables to provide tool-based tailoring support for IEC 61508: The tailoring universe would define a single 'SIL' property measured on a *ranks* scale from 1 to 4, with four FPVs $sil_{1..4}$ representing the SIL levels. Depending on the value of the 'SIL' property, and provided it is not unspecified, exactly one of the four variables would valuate to *true* and all others would valuate to *false*. We could represent the recommendation categories as constants valuating to according FTIs, e.g., [1, 1] for 'HR' and [0, 1] for '—'. Since IEC 61508 has no separate notions of necessity and feasibility, we would define identical necessitating and qualifying hypotheses per tailoring option, as follows:

$$hyp_n(o) = hyp_q(o) = (sil_1 \rightarrow c_1) \land (sil_2 \rightarrow c_2) \land (sil_3 \rightarrow c_3) \land (sil_4 \rightarrow c_4) \qquad (5.1)$$

with constants $c_{1...4} \in \{nr, -, r, hr\}$. For example, the condition for item 2 in Table 5.1 would look like this:

$$hyp_n(o) = hyp_q(o) = (sil_1 \rightarrow r) \land (sil_2 \rightarrow hr) \land (sil_3 \rightarrow hr) \land (sil_4 \rightarrow hr) \qquad (5.2)$$

Process Element	Process Element Example	Likely Tailorable Attribute	Alternatives	Considerations
Activity	Design Code Test Review Communicate	Frequency	Once/week	highly critical, mega or large
			Once/month	
			Once/quarter	
			At major milestones	
		Formality	Meeting w/ minutes	mega or large
			Informal meeting	large or medium
			Memo	large, medium, or small
			Email	medium or small
			Phone call	small
⋮	⋮	⋮	⋮	⋮

Table 5.2: Excerpt from a CMM tailoring table [GQ94, Table 4-1], with some added horizontal lines to clarify its structure.

Ginsberg/Qinn tailoring tables

A more elaborate table-based approach to process tailoring has been proposed by Ginsberg and Quinn in the context of CMM [GQ94]. They recommend that

> "the tailoring guidelines be developed by initially creating a table that indicates the process elements, the tailorable attributes for each element, the range for each attribute, and the considerations for selecting a particular range" [GQ94, p. 33].

According to this approach, it is not the CMM process elements, but the values for attributes of these process elements that are being tailored. Table 5.2 gives an example for tailoring guidelines about attributes of some activities. The leftmost column, somewhat confusingly labelled "process element," really identifies *types* of process elements, whereas the next column lists actual process elements. The third column identifies attributes applicable to the process elements, with permissible values in the fourth column. The last column states conditions for choosing specific values from the fourth column. The authors do not explain in great detail how to interpret this column, but it apparently expresses constraints on project properties. At least, the authors provide definitions for the size attributes as follows:

$$\text{small} = < 5 \text{ people}, < 6 \text{ months}, < \$200\text{K}, \text{ or } < 0.5 \text{ KSLOC}$$
$$\text{medium} = 5\text{–}15 \text{ people}, 0.5\text{–}1 \text{ year}, \text{ or } < \$0.5\text{M}, \text{ or } 0.5\text{–}5 \text{ KSLOC}$$
$$\text{large} = 10\text{–}25 \text{ people}, < 1\text{–}3 \text{ years}, < \$1\text{M}, \text{ or } 5\text{–}20 \text{ KSLOC}$$
$$\text{mega} > \text{large}$$

where amounts expressed in '$' apparently represent the budget of the project, and KSLOC stands for *thousand software lines of code* [Wikc].

At first glance, there seems to be a basic incompatibility with our approach to tailoring, since contrary to our definitions, Ginsberg's and Quinn's process elements are not atomic constituents of the process but are qualified by their attributes. Nevertheless, these different semantics do not alter the fact that the actual tailoring choices are of a binary nature: By treating the attribute values as tailoring options (column *alternatives*), we can translate the associated constraints to tailoring rules and only need to add one additional rule per tailorable attribute, stating that only one of the available alternatives may be chosen per attribute. In fact, the alternative values for attribute *formality* in Table 5.2 each determine a concrete technique that we might call a process element within our nomenclature; so the level at which tailoring occurs in Ginsberg's and Quinn's approach is not altogether that far away from our own.

Actually, our approach offers some advantages over the table-based approach discussed here. First, we can express tailoring rules that are much more complex than the limited space in tailoring tables could afford. Also, the concept of general tailoring rules introduced in Section 3.1.2 eliminates the limitation of having to express conditions separately for each tailoring option, and instead allows us to arbitrarily combine conditions about multiple options and measurements in single rules. Our separate notions of necessity and feasibility also make it possible to deal explicitly with contradictions in the tailoring context or between tailoring decisions, whereas the rules in Ginsberg's and Quinn's approach do not distinguish these two kinds of conditions. Finally, our application of fuzzy logic prevents the case that, for example, a tailoring option becomes eligible for a budget of $200K, but is disqualified at a budget of $201K, by softening these boundaries.

However, to come to a fair comparison of Ginsberg's and Quinn's approach and ours, we must acknowledge that Ginsberg and Quinn probably never intended their tailoring tables to be interpreted formally by a software tool, but by a human reader instead. For this purpose, their tailoring tables are an excellent solution, and a good trade-off between expressivity and manageability. In this light, their characterisations of size attributes described above are also sufficient, since a human reader will by himself bring in a certain amount of "fuzziness" when interpreting limits for budgets, durations, and so on.

Infosys

Infosys is an Indian IT consulting and offshore outsourcing company [Inf]. It has developed its own formal approach to tailoring OSSPs [Jal99, pp. 87–94] based on the Ginsberg/Quinn approach discussed above by refining tailoring to a two-step procedure:

> "Tailoring occurs at two levels: *Summary tailoring guidelines* suggest how some general activities should be performed in the project, based on some pro-

ject characteristics. *Detailed tailoring guidelines* list all activities in a process
for various life-cycle stages along with the information regarding tailoring for
each activity." [Jal99, p. 87]

The first step, summary tailoring, requires similar context characterisations as in Ginsberg's and Quinn's approach, including characteristics such as team size or application
criticality. However, instead of applying a context characterisation to choose individual
constituents for a process, they are applied to choose one or more *tables* of summary tailoring guidelines. In a second step, detailed tailoring, the selected summary guidelines are
applied in combination with detailed tailoring guidelines, expressed in a table format very
similar to Ginsberg's and Quinn's. The difference with regard to Ginsberg's and Quinn's
approach is that neither summary nor detailed tailoring guidelines are expressed formally,
but instead express conditions in informal language.

By applying Infosys' concept of two-step tailoring and at the same time maintaining Ginsberg's and Quinn's more formal approach to expressing conditions for isolated tailoring
options, we arrive at a close match with our tailoring approach for *ReqMan* as outlined in
Section 4.3. Instead of choosing values for process attributes, *ReqMan* requires choosing
suitable techniques to implement practices. Thus in both cases, we have similar 1-to-n
dependencies. The summary guidelines used at Infosys are similar to the tailorable hypotheses we put to use in Section 4.3.3. These hypotheses, too, provide a general direction
for tailoring, and interact with detailed tailoring guidelines when assessing tailoring decisions about *ReqMan* techniques. As in Ginsberg's and Quinn's approach, our *ReqMan*
example expresses all detailed tailoring guidelines as formal rules, while Infosys provides
formal rules only for selecting summary tailoring guidelines.

5.2.2 Tailoring by Example: Case-Based Approaches

We now present three examples of case-based approaches to process tailoring. Case-based
approaches differ fundamentally from our approach in that they do not derive tailoring
configurations analytically, but obtain complete configurations from a database. Therefore, case-based systems need not be aware of the structure of tailoring configurations.
Consequently, there is too little common ground for individual comparisons with our approach. We will instead provide a general discussion of the advantages and disadvantages
of rule-based and case-based tailoring approaches in Section 5.2.3.

TAME and the Experience Factory

The *Tailoring A Measured Environment* (TAME) project [BR87, BR88] proposed a
methodology for tailoring software processes "to the specific project goals and environment" [BR87], called the *Quality Improvement Paradigm* (QIP). The main focus of

TAME is not on the fine-grained tailoring of the core software processes, but on tailoring *supporting methods and tools* for process improvement. These may, but need not, be identical with methods and tools included in the core process model, but instead operate at a meta-level, examining the process itself to gain insight about its quality and productivity at all stages of the project.

At the core of the TAME project is a database containing process models, and supporting methods and tools for process improvement, along with a record of past experience with these items. This experience data is maintained in the form of characterisations of projects completed in the past. The database associates them with the process models, methods and tools applied in these projects, and with suggestions for better improvement methods and tools for similar future projects.

So at the level of actual process models, the tailoring support TAME offers is restricted to selecting a complete process model which has been successfully applied in a similar past project, while supporting methods and tools are offered as separate components. These do not only support ongoing projects, but also deliver the results required for a post-mortem analysis. Based on this final analysis, a newly finished project can be included in the database along with its characterisation, and with lessons learned in the form of *packages* containing an improved process model and better supporting methods and tools.

On the technical side, TAME obtains its recommendations by finding the closest match between the characterisation of a current project with characterisations of past projects in the database. This method is called *case-based reasoning* [Kol93, Ric00] and is common to many evidence-based Artificial Intelligence knowledge management systems. As the scope if TAME broadened, the TAME project was gradually superseded by the *Experience Factory* [ADH+01], which has maintained and refined the core concept of case-based reasoning introduced with TAME, but has extended its application to "experience management" of organisational knowledge.

Crystal

With his *Crystal Methodologies* [Coc02], Alistair Cockburn distinguishes different flavours of *process methodologies* that are colour-coded by the spectrum of the rainbow. The choice of methodology depends on three main factors—the size of the team, the criticality of the application domain, and the key objectives of the project. Within the three-dimensional space spanned by these factors, Cockburn identifies regions that each prescribe a specific process methodology. Cockburn defines the "weight" of a particular methodology as its cost in terms of resources required. Up to now, Cockburn has detailed only two light-weight variants of his family of methodologies [HRS04], *Crystal Clear* and *Crystal Orange*.

APTLY

A recent and thus far mostly unpublished tailoring approach is *Agile Process Tailoring and probLem analYsis* (APTLY), an ongoing doctoral dissertation project by Frank Keenan [Kee04]. APTLY will produce a *process knowledge base* with a general part containing "techniques for individual aspects of a process and patterns for complete processes [...] taken from the literature," and a specific part containing "organisational knowledge recorded as a set of process patterns for completed projects." Keenan plans to use case-based reasoning techniques to "store, retrieve, and update such patterns."

5.2.3 Comparison of Rule-Based and Case-Based Approaches

Rule-based tailoring approaches apply an inference mechanism to the context description on the grounds of formalised rules. Case-based tailoring approaches match the current context description against reference descriptions in a database and recommend a prefabricated tailoring configuration for the most similar reference case. Rule-based systems represent tailoring guidelines explicitly, whereas case-based systems represent them only implicitly in the associations between reference patterns and recommendations.

Case-based tailoring systems have the advantage that a learning strategy can be implemented easily by adding new tailoring patterns to the database after the completion of successful projects. Only limited formal expertise is required to enhance the database, since new experience is only added as evidence, and needs not be generalised to a more abstract representation. An additional advantage of this approach is that the contents of the database is inherently validated as long as only real-world cases are added to the database.

However, a system based on anecdotal information only gets useful after a certain number of cases have accumulated in the database. Assuming that few significant software projects take less than six months, and most of them take more than a year, a considerable amount of time will pass before enough tailoring experience will have been added to the database. Even large organisations which carry out more projects in parallel will have to put up with this initial delay, although the learning curve of the system will be much steeper later on. In most industrial applications of such a system, one will neither be able to remedy this by supplying a generic case base to start with, since it would have to be populated with experience from a similar environment, and other organisations are not likely to share their own case databases.

Another limitation of case-based approaches is the low granularity of tailoring recommendations. Since a case-based system contains no explicit guidelines about process tailoring, it can only provide recommendations in the form of one or a few prefabricated chunks making up a process model, which can then be adapted only manually in a trial-and-error fashion, to be appended to the database if the project was completed successfully. A case-based system is more of a recipe collection than a know-how reference; its recommendations cannot be questioned as is the case for rule-based systems with a justification component such as ours.

Even in the light of these practical difficulties, one might still advocate case-based approaches for their "soft" matching techniques that are liable to produce sensible solutions even when, in a similar situation, a "rigid" rule-based approach might disqualify all available alternatives because one condition was slightly missed. However, we have shown with our tailoring framework that FIL can bring similar softness to rule-based approaches.

The main disadvantage of the rule-based approach with regard to case-based systems is that expert knowledge about tailoring has to be translated to general, formal rules. Depending on the expressivity of the rule language and the breadth of the knowledge to be codified, this can become a complex task that requires good analytical skills. Not only is there a great effort involved to arrive at an initial configuration of the tailoring system, but the rule base also has to be extended manually if new experience needs to be added. If a rule system does not allow for the introduction of a certain degree of tolerance with regard to matching complex criteria and contradicting rules, there is also the danger of it provoking too much manual "fine-tuning," leading to too complex rules and bad maintainability.

Our approach has introduced the concepts of fuzziness and vagueness through FIL, and handles contradictions gracefully because, by virtue of the deontic operator, we can handle explicitly the case that a condition is necessary but unfeasible at the same time. Our experience from defining rules for the *ReqMan* project is that these characteristics of our approach result in only few unwanted dependencies between rules, and overall good maintainability. The initial analytical effort to create a rule base is further rewarded by its general applicability across different organisations.

Summing up, we conclude that the advantages of a rule-based tailoring system outweigh those of a case-based tailoring system in the majority of situations, and that our method of expressing rules effectively mitigates the drawbacks inherent to rule-based approaches.

5.3 Software-Assisted Tailoring: The V-Model XT

The *V-Model XT* [FRG04] is a standardised framework of development processes. It is the current official standard for IT systems projects in Germany, and is obligatory for all projects in the public sector.[3] It has been commissioned by the German government and

---

[3] Since most publications about the V-Model are in German, you might wish to refer to press release [IAB04] for a short introduction in English.

developed by a consortium of small and big companies, and of public and private research institutions. On its publication in late 2004 it replaced its predecessor, the *V-Model 97*, which had been the obligatory standard for publicly funded IT projects since 1997.

The suffix *XT* of the new V-Model stands for "extreme tailoring," and symbolises the unprecedented emphasis this process framework puts on process tailoring. Not only does it encourage process tailoring, it also supplies guidelines for tailoring and even provides limited tool support for tailoring an initial process description.

Since the V-Model XT is unique in its specificity about the tailoring procedure, we now take a more detailed look at tailoring in the V-Model XT. We will also show that the expressivity of our own TSS is sufficient of emulate the V-Model XT approach to tailoring, and we will express the V-Model XT tailoring framework in terms of a tailoring universe, tailoring options, and tailoring rules in Section 5.3.3.

Before that, in Section 5.3.1 we will first give a short overview of the key aspects of the V-Model XT process framework that are relevant to tailoring. Then we will outline the V-Model XT approach to tailoring in Section 5.3.2.

Due to the current lack of an official English version of the V-Model XT, in the following we introduce our own English translations of key terms and name the German equivalents in parentheses.

5.3.1 Key Aspects of the V-Model XT

The V-Model XT has a modular structure that is specified in one large XML file. The main components that make up the V-Model XT process framework consist of three *project types (Projekttypen)*, 18 *process modules (Vorgehensbausteine)*, and seven *project enactment strategies (Projektdurchführungsstrategien)*. We now discuss each of these kinds of components in turn.

Project Types

The available three project types each correspond to one of three scenarios:

1. system development project of a supplier

2. system development project of a contractor

3. introduction and maintenance of an organisation-specific process model

In case the organisation in question is both supplier and contractor, scenario 1 applies.

The choice of project types also has implications on available process modules and project enactment strategies; we will discuss these in Section 5.3.2 below.

Figure 5.1: UML model of the elements constituting a V-Model XT process module (simplified)

Figure 5.2: UML model of a V-Model XT project enactment strategy

Process Modules

The processes defined by the V-Model XT are product-centric. A *product* may be any kind of artefact such as a document, or even a physical object, that is applied, changed, or created in the course of a project. Every process module specifies a number of products. These products are associated with *roles* responsible for the product, and with *activities* affecting the product. Products may also depend upon the completion of other products (see Figure 5.1).

Project Enactment Strategies

A project enactment strategy defines *progress levels (Projektfortschrittsstufen)* that act as milestones or quality gates. While the project enactment strategy determines the sequence of the progress levels, their fulfilment is determined by the completion of sets of products with which individual progress levels are associated (see Figure 5.2).

5.3.2 Tailoring Support by the V-Model XT

The V-Model XT distinguishes *static* tailoring and *dynamic* tailoring. Static tailoring takes place at the start of each new project and results in a process manual that contains the process description. If necessary, the process description can be adapted in the course of the project, if required by circumstances. This is the dynamic variant of tailoring. We now examine both kinds of tailoring, starting with static tailoring.

Project feature	Possible values
Project objective	One of: • Introduction and maintenance of an organisation-specific process model • Embedded system • Hardware system • Complex system • Software system • System integration
Project role	One of: • Contractor • Supplier • Contractor and supplier
System life cycle phase	One of: • Development • Maintenance • Further development and migration
Commercial project management	yes/no
Quantitative project metrics	yes/no
COTS products	yes/no
User interface	yes/no
Safety and security	yes/no
High realisation risks	yes/no

Table 5.3: Project features used by the V-Model XT tailoring mechanism

Static Tailoring

Static tailoring is supported by a software application called the *project assistant (Projekt-assistent)*. The project assistant asks the user to characterise the project based on nine *project features (Projektmerkmale)*. Every project feature can be specified by the user by choosing an item from a list of possible answers. The V-Model XT project features are summarised in Table 5.3.

The project assistant allows the user to tailor the initial project description in two steps. In step one, it determines a project type based on the first two project features, *project objective* and *project role*. The user may alternatively choose the project type manually. Deciding on a project type also determines the tailoring status of each of the process modules and project enactment strategies. Every project type maps process modules to one of three categories: mandatory modules, optional modules, and prohibited modules. Mandatory process modules are always included in the process description, optional modules may be included in the second tailoring step, and prohibited modules cannot be included.

Likewise, the project type determines which project enactment strategies are available. In contrast with process modules, there are no obligatory project enactment strategies, so project enactment strategies will always have to be chosen in the second tailoring step.

Once the project type is chosen, step two of the static tailoring procedure determines which of the available process enactment strategies and optional process modules are to be included in the process description. The user can narrow down these choices by specifying one or more of the remaining seven project features, which, with a single exception, can all be answered simply with *yes* or *no*.

The V-Model XT does not encourage tailoring individual roles, products, or activities within process modules, or the adaptation of project enactment strategies. Neither does the project assistant support this kind of tailoring. The only way to perform this more fine-grained kind of tailoring would be to manually adapt the process description after it has been generated by the project assistant supplied with the V-Model XT.

Dynamic Tailoring

The notion of dynamic tailoring in the V-Model XT is restricted to choosing additional process modules in the course of the project. In order to maintain consistency with the state of the ongoing project, the addition of process modules is also restricted by *tailoring/product dependencies (Tailoring-Produktabhängigkeiten)*. V-Model XT specifies only five tailoring/product dependencies, and only as informal descriptions. Dynamic tailoring is not supported by the project assistant.

5.3.3 Tailoring the V-Model XT with our approach

The tailoring concept of the V-Model XT can be interpreted as a specialisation of the tailoring approach proposed in this work. We now illustrate this by modeling the V-Model XT tailoring concept with the means of our approach, defining tailoring options, a tailoring universe, and tailoring rules.

Tailoring Options

The tailoring options provided by the V-Model XT are the project types $t_i \in T$, the process modules $m_j \in M$, and the project enactment strategies $s_k \in S$:

$$O = T \cup M \cup S \tag{5.3}$$

Tailoring Universe

The tailoring universe consists of a single entity type *project*:

$$T = \{project\} \tag{5.4}$$

The project features in Table 5.3 can be expressed as properties for entity type *project*, with metrics measured on the 'name' scale as in Table 2.1 on page 14. The formal definitions of the properties and metrics associated with Table 5.3 are trivial so we do not detail them here.

We also introduce FPVs for use in the tailoring rules. Every FPV will valuate to one of only three values—*true* ($[1, 1]$), *false* ($[0, 0]$), or *unknown* ($[0, 1]$). We write each FPV in the form $F[V]$ where F represents a process feature from Table 5.3 and V represents a value for F. $F[V]$ valuates to *true* if value V is specified for the project feature F. If a value other than V is specified for project feature F, then variable $F[V]$ valuates to *false*. If project feature F is unspecified, every variable $F[V]$ valuates to *unknown*. For example, variable *user-interface[yes]* valuates to *true* if and only if *yes* has been specified for project feature *user interface*.

Tailoring Rules

Given the tailoring universe and tailoring options as defined above, we can now represent the tailoring rules of the V-Model XT in terms of rules for our TSS.

The XML file specifying the V-Model XT defines tailoring rules for project types, process modules, and project enactment strategies, i. e., for all components we have above identified as tailoring options in O. The rule format is the same for all of these components: For every component the V-Model XT maintains a list of project features paired with values. If a feature-value pair in such a list matches the actual value for that feature of the current project, then the associated component is to be included in the process description. We can represent the feature-value list for component $o \in O$ as

$$C_o = \{(f_1, v_1), \ldots, (f_n, v_n)\} \tag{5.5}$$

Above we have already introduced FPVs $f[v]$ that valuate to 'true' if and only if a feature f has the value v. With this we can express tailoring rules from the V-Model XT in deontic logic for every tailoring option o:[4]

$$O\left(\left(\bigvee_{(f,v)\in C_o} f[v]\right) \rightarrow \mathit{configured}(o)\right) \tag{5.6}$$

---

[4] The 'O' symbol in the rule represents the deontic operator "it is obliogatory that ... "; see Section 3.1.1 on pages 53 ff.

As described in Section 5.3.2, every process model also distinguishes mandatory, optional, and prohibited process modules, and specifies available process enactment strategies. These are also specified in the V-Model XT XML file. Therefore, for every process type t_i there are rules

$$O\left(configured(t_i) \rightarrow \bigwedge_{m \in M_i^m} configured(m) \right) \tag{5.7}$$

$$O\left(configured(t_i) \rightarrow \bigwedge_{m \in M_i^p} \neg configured(m) \right) \tag{5.8}$$

$$O\left(configured(t_i) \rightarrow \bigwedge_{s \in S_i^p} \neg configured(s) \right) \tag{5.9}$$

where $M_i^m \subseteq M$ are the process modules that are mandatory for process type t_i, and $M_i^p \subseteq M$ are the process modules that are prohibited for process type t_i, and $S_i^p \subseteq S$ are the project enactment strategies that are *not* available for process type t_i.

Up to here we have explained the tailoring rules that are defined explicitly in the XML definition file for the V-Model XT. However, experimentation with the project assistant included with the V-Model XT distribution reveals that more tailoring rules come to play than are provided in the XML definition of the V-Model XT. We have neither found mention of these additional tailoring rules in the official V-Model XT documentation, and have therefore reconstructed them empirically from the behaviour of the project assistant.

As the result of these observations, we have found that the V-Model XT project assistant seems to completely ignore tailoring rules in the form of (5.6) for deciding on the project type. Instead, it uses the following rules:

$$O(mutex(configured(t_1), configured(t_2), configured(t_3))) \tag{5.10a}$$

$$\begin{aligned} O(\ &if\text{-}then\text{-}else(project\text{-}objective[introduction\text{-}maintenance\text{-}ospm], \\ &\quad configured(t_3), \\ &\quad if\text{-}then\text{-}else(\ project\text{-}role[contractor], \\ &\quad\quad configured(t_2), \\ &\quad\quad configured(t_1)))) \end{aligned} \tag{5.10b}$$

where types $t_1 \ldots t_3$ correspond to the equally numbered types described in Section 5.3.1.

Another rule ensures that at least one project enactment strategy is selected:

$$O\left(\bigvee_{s \in S} configured(s) \right) \tag{5.11}$$

We have discovered no other irregularities in the behaviour of the project assistant that could not be reproduced with rules (5.10) and (5.11).

5.3.4 Enhancing V-Model XT Tailoring Assistance

We have shown that our TSS is able to emulate the tailoring approach of the V-Model XT. Granted the exceptions expressed in (5.10) and (5.11), it is surprising with how little expressivity the tailoring approach of the V-Model XT gets by. Equally, the adaptation of the V-Model XT tailoring approach to our TSS demonstrates the generality of our approach.

Furthermore, the greater expressivity of our TSS would allow us to provide useful extensions to the V-Model XT tailoring approach: We could even extend tool support to include dynamic tailoring, and provide formalised rules for tailoring/product dependencies. To achieve this, we would have to add V-Model XT products to the pool of tailoring options, and would possibly have to extend the tailoring universe by additional properties representing tailoring conditions expressed in the informal rules for tailoring/product dependencies provided by the V-Model XT.

In addition, we could introduce the notion of fuzziness to V-Model XT tailoring by introducing fuzzy variables that do not only valuate to one of the extremes *true*, *false*, and *unknown*, but to arbitrary fuzzy intervals. We could also extend the granularity of tailoring by allowing to tailor products, roles, and activities within process modules, by providing appropriate tailoring options and rules.

Since our TSS can import tailoring rules and universes from XML files, tailoring rules defined in the V-Model XT XML definition file could be converted and imported automatically, and could be kept in sync with future releases of the V-Model XT.

Our TSS also adds another benefit from the beginning—in contrast with the V-Model XT project assistant, the TSS can justify each of the tailoring decisions it encourages. Every justification ultimately traces back to the tailoring context, which in the case of the V-Model XT consists of the project features along with the values the user has specified for them. This added transparency helps the user understand the recommendations of the tailoring software, and increases his trust in the system.

5.4 Summary

In this chapter, we have argued that rule-based approaches to tailoring are better suited to many tailoring scenarios than case-based approaches. With regard to rule-based approaches, we have also underlined the universality of our approach by showing how it could be used to emulate other existing approaches. We have discussed the V-Model XT in particular detail since it is unique in its level of tool support for tailoring among process frameworks relevant in the software industry. Again we have demonstrated that our own approach is generic enough to emulate the V-Model XT approach, and provides greater flexibility for specifying tailoring rules and contexts.

These results provide evidence that our tailoring framework is fit for process tailoring in real-world software projects. While in our own work we have focused on the structure and methodology underlying this framework, future research should investigate techniques and concrete examples for practical applications of our approach. These could even include a wider range of uses, providing not only tailoring knowledge about software processes, but including a broader scope of organisational processes.

We have shown that prominent process models and process reference models do include tailoring guidelines. Due to their generally high complexity, tailoring these models is necessary in many cases. Even though available tailoring guidelines are sometimes vague or even only implicit, we could in every instance identify sources of information suitable for deriving tailoring rules, universes, and contexts within the framework of our approach.

6 Conclusion

6.1 Summary

Appropriate processes are processes that are adapted to a given environment such that they are as agile as possible, but as rigid as necessary. The act of adapting a process description to a given environment is called process tailoring. Among the existing approaches to process tailoring we have examined, there is no generic solution that allows to provide tailoring guidelines for tailoring appropriate processes, based on both experience knowledge and characteristics of the current project. The goal of our research was to fill this gap by laying the foundations for a software-based *tailoring support system* (TSS) that provides this kind of assistance for process tailoring. Consequently, we have put forward a framework for tailoring, have implemented it as a *Java* class library, and then have evaluated it in the context of a process model for requirements engineering.

Figure 6.1 represents the fundamental structure of our tailoring framework. It distinguishes two levels, one for abstract process models and one for concrete process descriptions. In the course of tailoring, a concrete process description is derived from an abstract process model. Both levels of the framework are divided into three aspects. Every aspect on the abstract level has a correspondent on the concrete level. The remainder of this summary follows this structure, starting with the three aspects of the abstract level.

On the abstract level of process models, our framework comprises three base components of tailoring guidelines, all of which need to be specified by the author of tailoring guidelines: A *tailoring universe* specifies types of entities relevant in tailoring situations, such as *project* or *client*, and properties for each type, such as *duration* or *experience*.

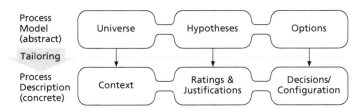

Figure 6.1: Schematic overview of our tailoring framework

Tailoring options represent the binary *yes/no* choices that have to be made in the course of tailoring—each tailoring option can be either included in a tailoring configuration, or excluded from it. For every tailoring option there are two *tailoring hypotheses*: one *necessitating hypothesis* that determines when choosing an option is required, and one *qualifying hypothesis* that determines when choosing an option is feasible. These tailoring hypotheses are in the form of logic propositions. For each tailoring option, they express dependencies with properties of entities, and with decisions about other tailoring options.

When tailoring a process model to a concrete process description, the user of a TSS specifies a *tailoring context* in terms of the entity types and their properties as specified in the tailoring universe. He may specify measurements or estimates for properties as intervals if a precise value is not known, or may even decide not to provide values for all properties or even not to instantiate some of the entity types. Then he tells the TSS to find the optimal tailoring configuration, that is, the configuration with the best rating. The TSS calculates the overall rating of a tailoring configuration from the ratings of all individual *tailoring decisions* about the tailoring options, which in turn it determines from the *fuzzy interval logic* (FIL) valuations of their respective hypotheses. Each rating is represented as an interval on a worst-best scale ranging from 0% to 100%, with the extremes of the rating intervals representing the pessimistic and the optimistic ratings. The rating of the tailoring configuration is derived from the optimistic ratings of all tailoring decisions. To find the tailoring configuration with the best rating, the TSS uses the T* optimisation algorithm which we have derived from the commonly known A* algorithm.

Specifying tailoring hypotheses directly is a difficult task because every hypothesis is focused exclusively on a single tailoring option. Expressing mutual dependencies between two or more tailoring options involves changing the hypotheses about *all* of these options. Therefore, to make defining tailoring hypotheses easier, we have introduced the concept of *general tailoring rules*, which do not express conditions about particular tailoring options, but instead formulate general restrictions about tailoring configurations. We have introduced deontic logic, a member of the family of modal logics, to derive a sound method of transforming a system of general tailoring rules to an equivalent system of tailoring hypotheses. This transformation method also has the added advantage of delivering hypotheses about *fuzzy propositional variables* (FPVs), making it possible to rate the congruency of properties of the tailoring context with a given tailoring configuration. Since values for properties are only supplied manually by the user and never suggested by the system, FPV ratings have no impact on the behaviour of the TSS, but can help to detect weaknesses and contradictions in the user's characterisation of the tailoring situation.

To provide transparency about the ratings calculated by the TSS, and to increase the user's trust in the ratings, we have devised algorithms that allow the TSS to supply justifications for all ratings. These justifications have a simple hierarchic structure of assertions where subordinate assertions express facts supporting their superordinate assertion. We have also proposed a way of rendering justification hierarchies so that they can be presented comprehensibly in a graphical user interface.

To give an example application of our tailoring framework, we have derived a tailorable process model from requirements management practices and techniques put forward by the *ReqMan* project, and have documented and discussed the optimal tailoring configuration calculated by our implementation of the TSS.

In a survey of related approaches to tailoring assistance we have argued that rule-based approaches such as ours have many advantages over case-based approaches, and have provided a detailed comparison with the V-Model XT, a recently established process model which, among the process models we have examined, is the first to include a software tool for tailoring assistance.

6.2 Validating our Theses

Now that we have briefly summarised the results of this work, we shall revisit our theses put forward in Section 1.2 to verify their fulfilment.

Thesis 1: Effective Tool Support for Process Tailoring

We have realised our tailoring framework as a *Java* class library that implements all data structures and algorithms put forward in the previous chapters, and which we applied successfully to calculate the example case discussed in Section 4.3 and exhibited in appendices B and C. We now seek to justify that a TSS based on this library can indeed provide appropriate support for process tailoring, as claimed by Thesis 1.

In our approach to process tailoring, tailoring decisions build on two sources of knowledge—short-term knowledge about the context and nature of the current project *(data)*, and long-term, general knowledge about processes *(experience)* that helps to judge the appropriateness of different practices and tools for the current project. In our tailoring framework (Section 2.2), short-term knowledge is expressed as the tailoring context in the form of measurements or estimates of metrics, whereas long-term knowledge is expressed as tailoring hypotheses. Based on this distinction, we can identify four basic types of tailoring decision-makers (Figure 6.2) with different deficits in current data, experience, or both. We will now look at the potential contributions of a TSS based on our framework to each of the four types of decision-makers introduced above.

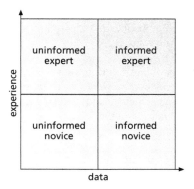

Figure 6.2: Requirements for tailoring assistance depend on user's short-term background (data) and long-term background (experience)

The Informed Expert: Focussing on Relevant Decisions

The act of tailoring usually involves going through a list of between 50 and 200 tailoring options, as is the case with the RUP or the V-Model XT, or with process assessment and improvement standards such as SPiCE or CMMI. In most cases the majority of these decisions is trivial and could easily be deduced from the context of the project, e. g., a project with a high security level would obviously have to include standard procedures to ensure corresponding security measures. While the experienced project manager does not depend on support for taking decisions, the tailoring procedure can be considerably streamlined if the underlying TSS anticipates all decisions that are sufficiently obvious. The project manager is presented with an overview of these anticipated decisions and, after having reviewed them, may then proceed to make the remaining, more substantial tailoring decisions.

The Informed Novice: Getting the Big Picture

An inexperienced project leader might be confronted with all necessary data but might not be able to interpret it completely and effectively. A TSS, by virtue of the entity types and properties defined in the tailoring universe, can help to single out relevant facts and estimates about the project from the available data. The TSS can then suggest tailoring decisions based on this information. Furthermore, it can check tailoring decisions made by the user for consistency and can uncover redundancies and gaps in the process description being tailored.

The Uninformed Expert: Managing Incomplete Information

By managing metrics that characterise key aspects of the current project, a TSS based on our framework can provide a structured overview of the project's context. The TSS thus facilitates the identification of the most relevant information deficits, i. e., those aspects that have not been sufficiently investigated, yet would significantly contribute to informed tailoring decisions. By analysing which properties of the tailoring context appear in the hypotheses about which tailoring options, a TSS can contribute significantly to the prioritisation of remaining information deficits.

The Uninformed Novice: Structuring the Project Initiation Phase

In the undesirable—but sometimes unavoidable—case that a project leader starts a new project with neither sufficient experience of similar projects nor the necessary data, the need for structured tailoring assistance is probably the most obvious. Not only will decisions need to be suggested and prioritised; first of all sufficient information will need to be collected to allow for well-founded decisions. As outlined above, a TSS based on our framework will be able to provide for both needs.

Appropriate Processes

In Section 1.1 we have argued that effective software development depends on *appropriate* processes, i. e., processes whose weight is optimally tuned to the requirements of the project. In providing the above explanations we have shown how tool-supported process tailoring supplies a suitable means to this end, provided that it takes into account both universal long-term knowledge about software processes, and short-term knowledge about the context of the current project.

Thesis 2: Adequate and Unambiguous Formal Language for Tailoring Guidelines

In Chapter 2 we have developed a tailoring framework that allows us to express tailoring guidelines formally in a three-part structure comprising available tailoring options, a tailoring universe, and tailoring hypotheses.

We can look at this structure as a formal language [Wikb], which, like most formal languages, is based on a set of primitive terms, and defines connectives serving to combine primitive terms to complex expressions.

The language we use is a specific instance of first-order logic [W⁺05a]: The primitives of our language are fuzzy propositional variables based on properties of the tailoring context ("the duration of the project is short"), and tailoring decisions ("option 'risk management' has been chosen"). The connectives of our language are the standard operators of classical logic [Sha05]. Tailoring hypotheses represent expressions in our language.

Unambiguousness

We have introduced formal semantics for obtaining interpretations of tailoring hypotheses by defining valuation functions for FPVs and for logic operators. The interpretation of any expression in our language takes the form of a FTI and is the result of recursively valuating all elements of the expression. Consequently, every tailoring hypothesis expressed in our language for tailoring guidelines has an unambiguous interpretation.

Adequacy

To justify that our language is also adequate for expressing tailoring guidelines, we shall briefly go into all three aspects of our language as introduced above.

Tailoring Options. Our tailoring model is based on the assumption that the act of tailoring consists in making a binary decision about each of a set of available options—each option can either be included in the tailoring configuration, or excluded from it. Although there are tailoring guidelines in literature that include choices with more than only two alternatives such as Ginsberg's and Quinn's approach (Section 5.2.1), we have discussed that these can be realistically reduced to equivalent binary options.

Tailoring Universe. Our concept of tailoring universes is based on a well-defined concept of metrics with clearly defined scale types (see Table 2.1 on page 14), allowing to measure or estimate properties of different types of entities. The distinction of entity types and instances enhances the degree of abstraction at which tailoring guidelines can be expressed, especially since it is possible to have more than one instance of any entity type, for example, for team members or subcontractors. Additional flexibility is granted by the ability to express measurements as ranges or even to leave them completely unspecified without obstructing the tailoring algorithm.

Tailoring Hypotheses. While expressions in first-order logic make it possible to formulate constraints on tailoring in an established way, the introduction of FPVs provides a simple but effective way of deriving intuitive statements from metrics, to serve as basic elements of tailoring hypotheses. This permits tailoring hypotheses to resemble conditional statements in informal language more closely.

These aspects of our tailoring framework, put to use in our tailoring example in Appendix B, document that the underlying formal language is indeed adequate for expressing unambiguous and adequate tailoring guidelines.

Thesis 3: Ordering Tailoring Configurations by Rating

As recapitulated above, every tailoring hypothesis can be interpreted in the form of a FTI. In Section 2.4.3 we have shown how ratings for tailoring decisions can be derived from the necessitating and qualifying hypotheses of their corresponding tailoring options. Then, in Section 2.4.4, we have shown how the overall rating of a tailoring configuration can be calculated from the ratings of its constituting tailoring decisions. The rating of every tailoring configuration consists of an interval within the range of 0% and 100%; we refer to the lower and upper boundaries of the interval as the pessimistic and optimistic ratings, respectively. As explained in Section 2.4.4, we use the optimistic ratings to weigh tailoring configurations against each other in terms of quality. We therefore have a total ordering relation between tailoring configurations and can consider Thesis 3 fulfilled.

Thesis 4: Efficient Detection of the Optimal Tailoring Configuration

As shown in Section 2.5, we were able to reduce the problem of tailoring optimisation to a shortest path search, restating the problem such as to fulfil all prerequisites imposed by the T* algorithm which we have derived from the A* algorithm. We have also introduced additional optimisations such as a heuristic to improve the order in which tailoring decisions are considered, and breaking the search space down into mutually independent partitions.

In Section 4.3.3, we have documented that calculating the optimal tailoring configuration for 41 options took 7.8 seconds on an average state-of-the-art personal computer. While an extensive empirical study of the algorithm's behaviour in other realistic tailoring scenarios is beyond the scope of this work, evidence from experiments similar to the example from Section 4.3 suggests that for less than 150 tailoring options the time required for finding the optimal configuration is roughly proportional to the number of tailoring options, permitting 150 tailoring options to be evaluated in approximately 30 seconds.

We therefore consider our implementation of the tailoring algorithm to support the claim put forward by Thesis 4.

Thesis 5: Maintainability and Scaleability

Thesis 5 demands maintainability of tailoring guidelines, even when they grow large and complex. As explained above, we can view the descriptive parts of our tailoring framework as a formal language for expressing tailoring guidelines. The maintainability of large and complex tailoring guidelines depends on at least two properties of this formal language—its expressiveness and its transparency. We now briefly explain both properties and then discuss their impact on the maintainability of tailoring guidelines in our tailoring framework.

Expressiveness

By the *expressiveness* of a language we understand its capabilities to interrelate its primitives to create new composite concepts.

We perceive two basic ways to expand the expressiveness of a language. First, introducing shorthand equivalents for commonly used complex expressions allows for expressing recurring patterns more tersely. Second, one can extend the underlying semantic apparatus, allowing to interrelate concepts of the language in entirely new ways.

While the first approach adds new *forms* of expressing concepts and is thus restricted to the syntactic level of the language, the second approach regards the semantic level of the language and extends the *scope* of concepts that can be expressed. Take, for example, a language that consists of the natural numbers and the operation of addition, with the usual semantics of integer arithmetic. A syntactic extension of the language would be to introduce a multiplication operator, all occurrences of which we could explain by a syntactic transformation—for example, 2×3 is equivalent to $3 + 3$ or $2 + 2 + 2$. However, a subtraction operator extends the semantics of our language since we cannot explain it by recurring to the notions available previously: We cannot rephrase an expression such as $5 - 2$ using the addition operation.

Transparency

We refer to the *transparency* of a language as the degree to which the formal interpretation of expressions in that language agrees with their intuitive readings. Since tailoring guidelines are expressed by humans, the language needs to be transparent such that evaluations of the guidelines conform with the author's expectations, even when the guidelines are complex.

Balancing Expressiveness and Transparency

The expressiveness and transparency of a language are interrelated, and the aim to optimise both has guided our design choices. We found transparency to be an inverse function of semantic complexity—the fewer semantic concepts are defined, the fewer are the possibilities of unanticipated interactions between those concepts. Therefore, we chose to use a language that permits every tailoring rule to be reduced to logic terms using only the operators of conjunction, disjunction, and negation.

Because we distinguish entity types and instances, we also needed to introduce quantifiers over multiple instances of an entity type, but we have restricted their scope to individual propositional variables (Section 2.4.6).

FPVs introduce common-sense notions about entities and their properties, mapping quantitative measurements to qualitative statements and thus eliminating the need to express numerical calculations within tailoring rules.

It was for reasons of transparency, too, that we have refrained from introducing other language elements, such as individual weightings for rules. While such weightings at first sight seem to be an attractive means for fine-tuning the behaviour of a rule-based system, we could not find a realisation that would not create unforeseen side-effects when applied broadly, and thus obscure the interactions between rules.

Where we extended the expressiveness of our language by a considerable degree was at the syntactic level, which not only makes expressing tailoring rules easier, but also increases their transparency if additional syntactic constructs reflect familiar notions from informal language. For example, the expression *if A then B else C* reads easier than $(\neg A \vee B) \wedge (A \vee C)$ even though they are equivalent by the definitions of our language for tailoring rules. As outlined in Section 3.1.3, we have introduced several such operators to assist in writing more legible tailoring rules.

We have achieved another significant improvement of expressiveness by introducing general tailoring rules in Section 3.1.2, which eliminate the need to express tailoring hypotheses in terms of the necessity or feasibility of single tailoring options, and instead allow for expressing arbitrary interrelations of tailoring decisions and propositional variables. These general rules are automatically transformed to the more restricted form of necessitating and qualifying hypotheses both about tailoring decisions and about properties of the tailoring context. Even though we have introduced deontic logic to provide a sound semantic foundation for this transformation, we hide deontic operators from the authors of rules and instead quietly wrap each rule with the deontic *obligation* operator O.

Given these considerations and our experience from modelling tailoring guidelines for the *ReqMan* project (Section 4.3), we have supplied both analytical and empirical arguments to support the claim that a tailoring system can indeed sustain the maintainability even of large and complex tailoring guidelines.

Thesis 6: Transparent Ratings of Tailoring Configurations

In Chapter 3.2 we have developed a mechanism to justify all tailoring decisions that make up a tailoring configuration. These justifications make it possible to identify the background of a particular configuration rating, by listing all the relevant facts contributing to that rating. These facts are presented in a simple causal hierarchy where the only relation is "*A* because *B*," while the structural complexity of the underlying rules remains hidden.

We have argued in Section 3.2.2 that this is in parallel with how we supply justifications in informal communication. The accepted way of substantiating a claim is to name supporting facts, implying that a causal relation exists without detailing its nature. We say "put your coat on because *it is raining*," and not "put your coat on because *it will prevent you from getting wet in the rain*."

The consequence of this approach is that the user of the TSS can process the justifications faster because he is given justifications in a familiar form, and because he does not have to analyse complex structures in order to understand the justifications: Instead, all information is given in the form of simple, independent assertions.

The use of FIL brings several advantages in this context. Fuzzy variables are closely related to approximate statements found in human reasoning and allow for the construction of intuitive informal statements about the tailoring context. Furthermore, fuzzy truth intervals express a degree of both applicability and certainty of a statement. Truth values that are close to 50% indicate that a statement is neither fully true nor fully refutable, whereas truth values close to either end of the scale from 0% to 100% expressly dismiss or acknowledge a statement. Likewise, the broader a truth interval gets, the lower is the certainty of the respective statement.

By providing a graphical representation of justifications and their constituting rated assertions as proposed in Section 3.2.5, the applicability and certainty of such assertions can be recognised visually and intuitively.

Appendix C provides an illustrative example of a justified tailoring configuration. This and the above considerations lead us to conclude that we have succeeded in devising a method for justifying the rating of a tailoring configuration in a transparent way, without requiring knowledge of the underlying rules.

6.3 Future Work

As is the case with all research problems, we, too, while finding solutions for the challenges we had set out to resolve, have uncovered new ones that we deem to be promising subjects for future research. We have encountered three main questions we consider worthwhile for further investigation of our tailoring framework: design criteria for tailoring guidelines, an appropriate user interface, and further fields of application.

Design Criteria for Tailoring Guidelines. We have introduced a language for expressing tailoring guidelines. As with any new language, one of the fundamental questions is in what way it can best be put to use. We have covered syntax and semantics of our language. We have not covered style. What style conventions and design criteria for tailoring guidelines help to best exploit the facilities of the language? How can tailoring rules be formulated and organised such as to maximise their clarity, maintainability, conciseness, and re-use potential? Further experience with real-world applications of our tailoring framework will supply the material and experience required to answer these questions.

Appropriate User Interface. What could an appropriate user interface to our TSS look like? This depends on how, and under which circumstances, our tailoring framework is to be put to use. In Section 3 we have identified two basic kinds of users—the process modeler, and the process tailorer. In Section 4.3.5 we have already argued that both kinds of users would benefit from an appropriate software tool that provides access to our framework, and we have outlined basic features of such a tool.

In Section 5.2.3 we have compared our rule-based approach with case-based approaches to tailoring. A software tool could enhance our rule-based approach by the advantages of case-based approaches by maintaining a database of previously tailored processes, and, with an adequate matching algorithm, could retrieve tailored processes from similar, successful past projects. Since in most cases a process tailorer will adapt and fine-tune a tailoring configuration recommended by the TSS (see Section 4.3.3), comparing his decisions with tailoring configurations from past projects could be a useful extension.

Another paradigm of the case-based approach is that of continually enriching the database with experience from new projects. To allow for the incorporation of new experience in a rule-based TSS, it is essential that tailoring rules are easily maintainable and understandable such that a larger community of experts are able, and encouraged, to regularly record their experiences in the rule base.

Further Fields of Application. Our work draws its motivation from practical problems of tailoring appropriate processes in the context of software engineering, and we have put forward an approach that is particularly suited to this end.

However, our core tailoring framework is based on a very general notion about tailoring, and makes no assumptions about the specific nature of software engineering or even processes: As long as a tailoring problem in any other domain structurally resembles the definitions of our tailoring framework—making an optimal choice of available options with regard to a context description—our formal tailoring framework can be applied to it equally well.

Consequently, our approach could also be applied to other kinds of processes such as the whole spectrum of systems engineering and management processes covered by

CMMI [Kaso4], the best practice processes for IT management put forward in the *IT Infrastructure Library* (ITIL) [ITI], or in fact any process that is documented formally.

The range of applications could be extended even further to accommodate for other problems such as

- planning purchases of configurable equipment for a specific purpose, where tailoring options represent optional components,

- configuring standardised products and services such as cars or cell phone contracts,

- putting together individual curricula of university courses, making sure that the selection of courses meets the minimum requirements for the targeted degree, or

- compiling exercise plans for members of a gym based on their age, gender, general health, and training goals.

6.4 Outlook

We believe that people often feel allergic towards processes because in many cases processes are not adapted to the specific requirements of a project. As a consequence, agility is often misinterpreted as the absence of clearly defined processes as opposed to the rigidity that is enforced by a defined process. However, the real choice is not whether to *have* a process or *not to have* a process, but the choice is whether to be *unconscious* of it or whether to shape it *explicitly*. Software-assisted tailoring is a powerful tool to shape a process model to the specific needs of a project. We expect it not only to make tailoring easier, but also to contribute to make tailoring more common practice. Tailoring is essential to overcoming the sharp divide between agility and rigidity because it aims at always getting the process that is as agile as possible, but as rigid as necessary. We therefore hope that our work will contribute to strengthening the role of tailoring in software engineering practice.

Appendix

A Mathematical Notation

Type	Notation	Remark
Set of Boolean truth values	$\mathbb{B}$	$\mathbb{B} = \{\text{true}, \text{false}\}$
Set of natural numbers	$\mathbb{N}$	$1, 2, 3, \ldots$
Set of integers	$\mathbb{Z}$	$\ldots, -3, -2, -1, 0, 1, 2, 3, \ldots$
Set of whole numbers	$\mathbb{Z}^*$	$0, 1, 2, 3, \ldots$
Set of real numbers	$\mathbb{R}$	
Empty set	$\emptyset$	
Constant	$a, b, \ldots$	regular lower case letters
Variable	$a, b, \ldots$	italicised lower case letters
Set	$\mathbf{A}, \mathbf{B}, \ldots$	bold capitals
Set constant	$A, B, \ldots$	capitals
Set variable	$A, B, \ldots$	italicised capitals
Power set*	$\mathcal{P}(\mathbf{A})$	
Function name	$f, g, \ldots$	italicised lower case letters
Function/mapping†	$f : \mathbf{A} \mapsto \mathbf{B}$	
Operator name	$\min, \max, \ldots$	regular lower case letters
Structure/tuple	$\mathbf{A}, \mathbf{B}, \ldots$	bold capitals
Set of all mappings from $\mathbf{A}$ to $\mathbf{B}$	$\mathbf{B}^{\mathbf{A}}$	
Equivalence relation	$\equiv$	

---

*The set of all possible subsets of $\mathbf{A}$.
†The terms *function* and *mapping* are equivalent.

A Mathematical Notation

B *ReqMan*Tailoring Guidelines

B.1 Tailoring Universe: Entity Types and Properties

Project

- project duration

  Scale: Amount (months)

Variable	Definition
– the duration is short	*Ramp:* 12–6 months

- team size

  Scale: Count (persons)

Variable	Definition
– the team is small	*Ramp:* 6–2 persons

- criticality of the application domain

  Scale: Grades

  1. insubstantial
  2. well below project cost
  3. as project cost
  4. well above project cost
  5. human injuries
  6. human casualties

Variable	Definition
– application domain is critical	*Ramp:* 1–5 [insubstantial – human injuries]

- product type

Scale: One of

- GUI
- Embedded
- COTS

Variable	Definition
– product is a GUI application	*Equals:* 1 [GUI]
– product is an embedded application	*Equals:* 2 [Embedded]
– product is a COTS application	*Equals:* 3 [COTS]

- **expected number of features**

Scale: Count (Features)

Variable	Definition
– many features have to be realised	*Ramp:* 10–30 Features

- **documentation relevant to the project is available**

Scale: Grades

1. false
2. mostly false
3. partially true
4. mostly true
5. true

Variable	Definition
– documentation relevant to the project is available	*Ramp:* 1–5 [false – true]

Stakeholder (abstract)

- **innovativeness**

Scale: Grades

1. unacceptable
2. bad
3. average
4. good
5. excellent

Variable	Definition
– possesses good visionary skills	*Ramp:* 2.5–4.5 [average – good]

- **formal competency**

  Scale: Grades

  1. none
  2. can understand formal structures if they are documented
  3. can gain an understanding of unfamiliar formal structures
  4. understands formal structures intuitively
  5. can create formal structures with some assistance
  6. can create formal structures independently

Variable	Definition
– has passive formal competency	*Ramp:* 1–4 [none – understands formal structures intuitively]
– has active formal competency	*Ramp:* 4–6 [understands formal structures intuitively – can create formal structures independently]

Internal Stakeholder (abstract)

Inherits: Stakeholder

- **work experience from similar projects**

  Scale: Amount (person years)

Variable	Definition
– is a beginner	*Ramp:* 1–0.5 person years
– has basic experience	*Ramp:* 0.5–1 person years
– is experienced	*Ramp:* 1–2 person years
– is very experienced	*Ramp:* 2–4 person years

External Stakeholder (abstract)

Inherits: Stakeholder

- **reachability**

  Scale: Grades

1. not directly reachable
2. by phone on appointment
3. always by phone
4. personally by appointment
5. personally on my site (by appointment)
6. always on my site

Variable	Definition
– is always available on my site	*Equals:* 6 [always on my site]
– is available on my site by appointment	*At least:* 5 [personally on my site (by appointment)]
– is available for meetings	*At least:* 4 [personally by appointment]
– is available by phone	*At least:* 3 [always by phone]
– is available by phone on appointment	*At least:* 2 [by phone on appointment]
– is available for discussions	*Ramp:* 0.5–3.5 [not directly reachable – always by phone]

- **size**

Scale: Count (persons)

Variable	Definition
– the group is big enough for workshops	*Ramp:* 0–3 persons
– the group is small enough for workshops	*Ramp:* 20–10 persons

- **conflicts between stakeholders**

Scale: Grades

1. false
2. mostly false
3. partially true
4. mostly true
5. true

Variable	Definition
– there are conflicts with other stakeholders	*Ramp:* 1–5 [false – true]

Team Member

Inherits: Internal Stakeholder

- **experience with requirements management**

  Scale: Amount (person years)

Variable	Definition
– has basic experience with requirements management	*Ramp:* 0.5–2 person years

- **prior knowledge of the application domain**

  Scale: Amount (person years)

Variable	Definition
– has basic knowledge of the application domain	*Ramp:* 0.5–2 person years

Client

Inherits: External Stakeholder

User Group

Inherits: External Stakeholder

- **access to application domain**

  Scale: Grades

  1. false
  2. mostly false
  3. partially true
  4. mostly true
  5. true

Variable	Definition
– can be visited in their surroundings	*Ramp:* 1–5 [false – true]

B.2 Tailoring Options

- Practice: Elicit Functional Requirements [ODKE05, O$^+$05]

- Practice: Elicit Non-Functional Requirements [ODKE05, O$^+$05]

- Practice: Determine Scope [ODKE05, O$^+$05]

- Practice: Integrate Stakeholders [ODKE05, O$^+$05]

- Practice: Identify Stakeholders and Sources [ODKE05, O$^+$05]

- Practice: Elicit Goals [ODKE05, O$^+$05]

- Practice: Elicit Tasks and Business Process [ODKE05, O$^+$05]

- Use Cases [Lau02]

- MisUse Cases [SO01]

- BPMN [BPM05]

- EPC Modelling [Sch99]

- Quality Function Deployment [Lau02]

- Goal-Domain Analysis [Lau02]

- Cost-Benefit Analysis [Lau02]

- On-Site Customer [Bec01]

- Pseudo-On-Site Customer [JAH00]

- Document Studies [Lau02]

- Stakeholder Analysis [Lau02]

- Lauesen Group Interview Technique [Lau02]

- Lauesen Interview Technique [Lau02]

- User Observation [Lau02]

- Task Demonstration [Lau02]

- Questionnaires [Lau02]

- Brainstorming [Lau02]

- Focus Groups [Lau02]

- Domain Workshops [Lau02]

- Design Workshops [Lau02]

- Features from Task Descriptions [Lau02]

- Tasks & Support [Lau02]

- Task-Feature Matrix [Lau02]

- Prototyping [Lau02]

- Pilot Experiments [Lau02]

- Study Similar Companies [Lau02]

- Analysis of Project Risks [Lau02]

- Ask Suppliers to Obtain Ideas about Useful Features [Lau02]

- Negotiation Technique: Mutual Stakeholder Analysis [Lau02]

- Negotiation Technique: Moderated Group Discussion with Conflict Resolution [Lau02]

- Risk Workshop [HHMS04]

- Hypothesis: Advanced elicitation practices should be carried out.

- Hypothesis: Risks and consequences must be elicited

- Hypothesis: Ensure completeness of elicitation [Lau02]

B.3 Tailoring Rules

It should be the case that:

1. **if** *Project*: application domain is critical

   then it should be the case that **chosen:** Hypothesis: Advanced elicitation practices should be carried out.

2. **if chosen:** Hypothesis: Advanced elicitation practices should be carried out.

   then it should be the case that **an arbitrary** *Team Member*: has basic experience with requirements management

3. **all of:**

   - **chosen:** Practice: Elicit Functional Requirements
   - **chosen:** Practice: Integrate Stakeholders
   - **chosen:** Practice: Identify Stakeholders and Sources

4. **if at least one** of: **chosen:** Hypothesis: Advanced elicitation practices should be carried out.

   then it should be the case that **all** of:

   - **chosen:** Practice: Elicit Non-Functional Requirements
   - **chosen:** Practice: Determine Scope
   - **chosen:** Practice: Elicit Goals

5. **if all** of:

   - **chosen:** Hypothesis: Advanced elicitation practices should be carried out.
   - **at least one** of:
     - *Project*: product is a GUI application
     - *Project*: product is a COTS application

   then it should be the case that **chosen:** Practice: Elicit Tasks and Business Process

6. **if chosen:** Practice: Elicit Tasks and Business Process

   then it should be the case that **all** of:

   - **at least one** of:
     - **chosen:** EPC Modelling
     - **chosen:** BPMN
     - **chosen:** Use Cases
     - **chosen:** Stakeholder Analysis
     - **chosen:** Lauesen Interview Technique
     - **chosen:** Lauesen Group Interview Technique
     - **chosen:** User Observation
     - **chosen:** Task Demonstration
     - **chosen:** Brainstorming
     - **chosen:** Focus Groups
     - **chosen:** Domain Workshops
     - **chosen:** Design Workshops
     - **chosen:** Prototyping
     - **chosen:** Ask Suppliers to Obtain Ideas about Useful Features
   - **if every** *External Stakeholder*: there are conflicts with other stakeholders
     then it should be the case that **at least one** of:
     - **chosen:** Negotiation Technique: Mutual Stakeholder Analysis
     - **chosen:** Negotiation Technique: Moderated Group Discussion with Conflict Resolution
   - **if** *Project*: application domain is critical
     then it should be the case that **chosen:** Risk Workshop
   - **if** *Project*: documentation relevant to the project is available
     then it should be the case that **chosen:** Document Studies

7. **if chosen:** Practice: Elicit Functional Requirements

   then it should be the case that **at least one** of:

   - **chosen:** Use Cases
   - **chosen:** MisUse Cases
   - **chosen:** Lauesen Interview Technique
   - **chosen:** Lauesen Group Interview Technique
   - **chosen:** Focus Groups
   - **chosen:** Domain Workshops
   - **chosen:** Design Workshops
   - **chosen:** Prototyping
   - **chosen:** Pilot Experiments
   - **chosen:** Study Similar Companies
   - **chosen:** Ask Suppliers to Obtain Ideas about Useful Features
   - **chosen:** Task-Feature Matrix
   - **chosen:** Features from Task Descriptions
   - **chosen:** Tasks & Support

8. **if chosen:** Practice: Elicit Non-Functional Requirements

   then it should be the case that **at least one** of:

   - **chosen:** Prototyping
   - **chosen:** Pilot Experiments
   - **chosen:** Study Similar Companies
   - **chosen:** Ask Suppliers to Obtain Ideas about Useful Features
   - **chosen:** Task-Feature Matrix
   - **chosen:** Features from Task Descriptions
   - **chosen:** Tasks & Support

9. **if chosen:** Practice: Determine Scope

   then it should be the case that

   > **if all** of:
   >
   > - *Project*: many features have to be realised
   > - *Project* **does not fulfill:** the duration is short
   >
   > **then** it should be the case that **at least one** of:
   >
   > - **chosen:** Quality Function Deployment
   > - **chosen:** Goal-Domain Analysis
   >
   > **else** it should be the case that **at least one** of:
   >
   > - **chosen:** Cost-Benefit Analysis
   > - **chosen:** Study Similar Companies

- **chosen:** Risk Workshop

10. **if chosen:** Practice: Integrate Stakeholders

    then it should be the case that **all** of:

    - **chosen:** Practice: Identify Stakeholders and Sources
    - **at least one** of:
      - **chosen:** On-Site Customer
      - **chosen:** Pseudo-On-Site Customer

11. **if chosen:** Practice: Identify Stakeholders and Sources

    then it should be the case that **at least one** of:

    - **chosen:** Stakeholder Analysis
    - **chosen:** Focus Groups
    - **chosen:** Domain Workshops
    - **chosen:** Design Workshops

12. **if chosen:** Practice: Elicit Goals

    then it should be the case that **at least one** of:

    - **chosen:** Quality Function Deployment
    - **chosen:** Goal-Domain Analysis
    - **chosen:** Lauesen Interview Technique
    - **chosen:** Lauesen Group Interview Technique
    - **chosen:** User Observation
    - **chosen:** Task Demonstration
    - **chosen:** Focus Groups
    - **chosen:** Domain Workshops
    - **chosen:** Design Workshops

13. **if at least one** of:

    - **chosen:** EPC Modelling
    - **chosen:** BPMN
    - **chosen:** Use Cases
    - **chosen:** MisUse Cases
    - **chosen:** Quality Function Deployment
    - **chosen:** Goal-Domain Analysis

    then it should be the case that **all** of:

    - **an arbitrary** *Team Member*: has active formal competency
    - *Client*: has passive formal competency

14. **if chosen:** MisUse Cases

    then it should be the case that **chosen:** Use Cases

15. **if chosen:** Quality Function Deployment

    then it should be the case that **all** of:

    - *Project*: the team is small
    - *Project* **does not fulfill:** the duration is short

16. **if chosen:** On-Site Customer

    then it should be the case that *Client*: is always available on my site

17. **if chosen:** Document Studies

    then it should be the case that *Project*: documentation relevant to the project is available

18. **if at least one** of:

    - **chosen:** User Observation
    - **chosen:** Task Demonstration

    then it should be the case that **all** of:

    - **every** *User Group*: can be visited in their surroundings
    - **every** *User Group*: is available by phone on appointment

19. **if chosen:** Brainstorming

    then it should be the case that **all** of:

    - **an arbitrary** *External Stakeholder*: possesses good visionary skills
    - **an arbitrary** *Team Member*: possesses good visionary skills

20. **if chosen:** Questionnaires

    then it should be the case that **all** of:

    - **an arbitrary** *Team Member*: has basic knowledge of the application domain
    - **no** *External Stakeholder*: the group is small enough for workshops
    - **at least one** of:
        - **chosen:** Practice: Elicit Tasks and Business Process
        - **chosen:** Practice: Elicit Goals
        - **chosen:** Practice: Identify Stakeholders and Sources
        - **chosen:** Practice: Integrate Stakeholders

21. **if at least one** of:

    - **chosen:** Focus Groups

- **chosen:** Domain Workshops
- **chosen:** Design Workshops

then it should be the case that **all of:**

- **every** *User Group*: the group is big enough for workshops
- **every** *User Group*: is available for meetings

22. **if at least one** of:

    - **chosen:** Domain Workshops
    - **chosen:** Design Workshops

    then it should be the case that **all of: every** *User Group*: the group is small enough for workshops

23. **if chosen:** Design Workshops

    then it should be the case that *Project*: product is a GUI application

24. **if every** *User Group*: can be visited in their surroundings

    then it should be the case that **every** *User Group*: is available by phone on appointment

25. **all of:**

    - **if** *Project*: product is a COTS application
      then it should be the case that **chosen:** Pilot Experiments
    - **if chosen:** Pilot Experiments
      then it should be the case that **every** *User Group*: is available for discussions

26. **if chosen:** Study Similar Companies

    then it should be the case that *Project* **does not fulfill:** many features have to be realised

27. **no** *External Stakeholder*: there are conflicts with other stakeholders

28. **not all of:**

    - **chosen:** EPC Modelling
    - **chosen:** BPMN

29. **not all of:**

    - **chosen:** Goal-Domain Analysis
    - **chosen:** Quality Function Deployment

30. **not all of:**

    - **chosen:** Pseudo-On-Site Customer

- **chosen:** On-Site Customer

31. **if chosen:** Hypothesis: Risks and consequences must be elicited

    then it should be the case that **at least one** of:

    - **chosen:** Domain Workshops
    - **chosen:** Prototyping
    - **chosen:** Pilot Experiments
    - **chosen:** Study Similar Companies
    - **chosen:** Ask Suppliers to Obtain Ideas about Useful Features
    - **chosen:** Risk Workshop
    - **chosen:** Analysis of Project Risks
    - **chosen:** Cost-Benefit Analysis
    - **chosen:** Goal-Domain Analysis

32. **if all** of:

    - *Project*: the duration is short
    - *Project*: many features have to be realised

    then it should be the case that **chosen:** Hypothesis: Risks and consequences must be elicited

33. **if chosen:** Hypothesis: Ensure completeness of elicitation

    then it should be the case that **at least one** of:

    - **chosen:** Goal-Domain Analysis
    - **chosen:** User Observation
    - **chosen:** Task Demonstration
    - **chosen:** Document Studies
    - **chosen:** Prototyping
    - **chosen:** Pilot Experiments

34. **chosen:** Hypothesis: Ensure completeness of elicitation

C ReqMan Sample Tailoring Recommendation

C.1 Tailoring Context

Client "BigCorp"

Property	Scale	Measurement
conflicts between stakeholders	*Grades* 1. false 2. mostly false 3. partially true 4. mostly true 5. true	3–4 [partially true – mostly true]
formal competency	*Grades* 1. none 2. can understand formal structures if they are documented 3. can gain an understanding of unfamiliar formal structures 4. understands formal structures intuitively 5. can create formal structures with some assistance 6. can create formal structures independently	1–3 [none – can gain an understanding of unfamiliar formal structures]
innovativeness	*Grades* 1. unacceptable 2. bad 3. average 4. good 5. excellent	3–4 [average – good]

Client "BigCorp" (contd.)

Property	Scale	Measurement
reachability	*Grades* 1. not directly reachable 2. by phone on appointment 3. always by phone 4. personally by appointment 5. personally on my site (by appointment) 6. always on my site	3 [always by phone]

Project "Alpha"

Property	Scale	Measurement
criticality of the application domain	*Grades* 1. insubstantial 2. well below project cost 3. as project cost 4. well above project cost 5. human injuries 6. human casualties	1–2 [insubstantial – well below project cost]
project duration	*Amount (months)*	6.0–9.0 months
documentation relevant to the project is available	*Grades* 1. false 2. mostly false 3. partially true 4. mostly true 5. true	2–4 [mostly false – mostly true]
expected number of features	*Count (Features)*	25–50 Features
team size	*Count (persons)*	3–5 persons
product type	*One of* • GUI • Embedded • COTS	1 [GUI]

Team Member "Elli Citator"

Property	Scale	Measurement
formal competency	*Grades* 1. none 2. can understand formal structures if they are documented 3. can gain an understanding of unfamiliar formal structures 4. understands formal structures intuitively 5. can create formal structures with some assistance 6. can create formal structures independently	3–4 [can gain an understanding of unfamiliar formal structures – understands formal structures intuitively]
innovativeness	*Grades* 1. unacceptable 2. bad 3. average 4. good 5. excellent	2–4 [bad – good]
experience with requirements management	*Amount (person years)*	1.2 person years
work experience from similar projects	*Amount (person years)*	1.5 person years

Team Member "Dave Lopper"

Property	Scale	Measurement
innovativeness	*Grades* 1. unacceptable 2. bad 3. average 4. good 5. excellent	4–5 [good – excellent]
experience with requirements management	*Amount (person years)*	0.5 person years

Property	Scale	Measurement
work experience from similar projects	*Amount (person years)*	3.0 person years
formal competency	*Grades* 1. none 2. can understand formal structures if they are documented 3. can gain an understanding of unfamiliar formal structures 4. understands formal structures intuitively 5. can create formal structures with some assistance 6. can create formal structures independently	6 [can create formal structures independently]

User Group "UseWorks"

Property	Scale	Measurement
size	*Count (persons)*	10–15 persons
reachability	*Grades* 1. not directly reachable 2. by phone on appointment 3. always by phone 4. personally by appointment 5. personally on my site (by appointment) 6. always on my site	2–3 [by phone on appointment – always by phone]
innovativeness	*Grades* 1. unacceptable 2. bad 3. average 4. good 5. excellent	1–2 [unacceptable – bad]

Property	Scale	Measurement
conflicts between stakeholders	*Grades* 1. false 2. mostly false 3. partially true 4. mostly true 5. true	2 [mostly false]
formal competency	*Grades* 1. none 2. can understand formal structures if they are documented 3. can gain an understanding of unfamiliar formal structures 4. understands formal structures intuitively 5. can create formal structures with some assistance 6. can create formal structures independently	1 [none]

C.2 Tailoring Decisions

C.2.1 Overview

● ✓ **recommended:** Practice: Elicit Functional Requirements . . . . [▭▮]

● ✓ **recommended:** Practice: Elicit Non-Functional Requirements . . . . [▭▮]

● ✓ **recommended:** Practice: Determine Scope . . . . [▭█]

● ✓ **recommended:** Practice: Integrate Stakeholders . . . . [▭▮]

● ✓ **recommended:** Practice: Identify Stakeholders and Sources . . . . [▭▮]

● ✓ **recommended:** Practice: Elicit Goals . . . . [▭▮]

○ ✓ **not recommended:** Practice: Elicit Tasks and Business Process . . . . [▭▮]

○ ✓ **not recommended:** Use Cases . . . . [▭▮]

○ ✓ **not recommended:** MisUse Cases . . . . [▭▮]

○ ✓ **not recommended:** BPMN . . . . [▭▮]

○ ✓ **not recommended**: EPC Modelling.................................

○ (✓) **not recommended**: Quality Function Deployment...................

○ (✓) **not recommended**: Goal-Domain Analysis.........................

● (✓) **recommended**: Cost-Benefit Analysis..............................

○ ✓ **not recommended**: On-Site Customer.............................

● ✓ **recommended**: Pseudo-On-Site Customer...........................

○ ✓ **not recommended**: Document Studies..............................

● ✓ **recommended**: Stakeholder Analysis................................

● ✓ **recommended**: Lauesen Group Interview Technique..................

○ ✓ **not recommended**: Lauesen Interview Technique....................

○ ✓ **not recommended**: User Observation..............................

○ ✓ **not recommended**: Task Demonstration............................

○ ✓ **not recommended**: Questionnaires................................

○ ✓ **not recommended**: Brainstorming.................................

○ ✓ **not recommended**: Focus Groups..................................

○ ✓ **not recommended**: Domain Workshops.............................

○ ✓ **not recommended**: Design Workshops.............................

○ ✓ **not recommended**: Features from Task Descriptions.................

○ ✓ **not recommended**: Tasks & Support...............................

○ ✓ **not recommended**: Task-Feature Matrix...........................

● ✓ **recommended**: Prototyping...

○ ✓ **not recommended**: Pilot Experiments.............................

○ ✓ **not recommended**: Study Similar Companies.......................

○ ✓ **not recommended**: Analysis of Project Risks.......................

○ ✓ **not recommended**: Ask Suppliers to Obtain Ideas about Useful Features..

○ ✓ **not recommended**: Negotiation Technique: Mutual Stakeholder Analysis...

○ ✓ **not recommended:** Negotiation Technique: Moderated Group Discussion with Conflict Resolution.......................................▢▮

○ ✓ **not recommended:** Risk Workshop ....▢▮

● ✗ **[recommended:** Hypothesis: Advanced elicitation practices should be carried out.]...▮▢

● (✓) **recommended:** Hypothesis: Risks and consequences must be elicited ..▢▮

● ✓ **recommended:** Hypothesis: Ensure completeness of elicitation ....▢▮

C.2.2 Justifications

● ✓ **recommended:** Practice: Elicit Functional Requirements ....▢▮

   ✓ this option is neccessary ....▢▮

      • always

   ✓ this option is feasible (after applying feasibility offset 0.2) ....▢▮

     ✓ this option is feasible......................................▢▮

       ✓ chosen: Prototyping....................................▢▮

       ✓ chosen: Lauesen Group Interview Technique ....▢▮

● ✓ **recommended:** Practice: Elicit Non-Functional Requirements ....▢▮

   ✓ this option is neccessary ....▢▮

     ✓ chosen: Hypothesis: Advanced elicitation practices should be carried out..▢▮

   ✓ this option is feasible (after applying feasibility offset 0.2) ....▢▮

     ✓ this option is feasible......................................▢▮

       ✓ chosen: Prototyping....................................▢▮

● ✓ **recommended:** Practice: Determine Scope ....▢▮

   ✓ this option is neccessary ....▢▮

     ✓ chosen: Hypothesis: Advanced elicitation practices should be carried out..▢▮

   ✓ this option is feasible (after applying feasibility offset 0.2) ....▢▮

     (✓) this option is feasible.....................................▢▮

       ✓ chosen: Cost-Benefit Analysis............................▢▮

       (✓) Project: the duration is short ....▢▮

         (✓) Alpha: the duration is short ....▢▮

           • project duration=[6.0 months–9.0 months]

● ✓ **recommended:** Practice: Integrate Stakeholders .... ☐

   ✓ this option is neccessary .... ☐

      • always

   ✓ this option is feasible (after applying feasibility offset 0.2) .... ☐

      ✓ this option is feasible .... ☐

         ✓ chosen: Pseudo-On-Site Customer .... ☐

         ✓ chosen: Practice: Identify Stakeholders and Sources ...... ☐

● ✓ **recommended:** Practice: Identify Stakeholders and Sources .... ☐

   ✓ this option is neccessary .... ☐

      • always

   ✓ this option is feasible (after applying feasibility offset 0.2) .... ☐

      ✓ this option is feasible .... ☐

         ✓ chosen: Stakeholder Analysis .... ☐

● ✓ **recommended:** Practice: Elicit Goals .... ☐

   ✓ this option is neccessary .... ☐

      ✓ chosen: Hypothesis: Advanced elicitation practices should be
carried out. .... ☐

   ✓ this option is feasible (after applying feasibility offset 0.2) .... ☐

      ✓ this option is feasible .... ☐

         ✓ chosen: Lauesen Group Interview Technique .... ☐

○ ✓ **not recommended:** Practice: Elicit Tasks and Business Process .... ☐

   ✗ [this option is neccessary] .... ☐

      ✗ [excluded: Practice: Integrate Stakeholders] .... ☐

      ✗ [excluded: Practice: Elicit Goals] .... ☐

      ✗ [chosen: Questionnaires] .... ☐

      ✗ [Project: product is a COTS application] .... ☐

         ✗ [Alpha: product is a COTS application] .... ☐

            • product type=GUI

      ✗ [Project: product is a GUI application] .... ☐

         ✗ [Alpha: product is a GUI application] .... ☐

            • product type=GUI

      ✗ [excluded: Practice: Identify Stakeholders and Sources] .... ☐

○ ✓ **not recommended:** Use Cases .... ▭▬

  ✗ [this option is neccessary] .... ▭▬

    ✗ [excluded: Lauesen Group Interview Technique] .... ▭▭

    ✗ [chosen: Practice: Elicit Tasks and Business Process] .... ▭▭

    ✗ [excluded: Stakeholder Analysis] .... ▭▭

    ✗ [chosen: MisUse Cases] .... ▭▭

    ✗ [excluded: Prototyping] .... ▭▭

  ▲ (this option is feasible) .... ▮▭

    ▲ (Client: has passive formal competency) .... ▮▭

      ▲ (BigCorp: has passive formal competency) .... ▮▭

        • formal competency=[none–can gain
         an understanding of unfamiliar formal structures]

    ▲ (Team Member [any 1 entity]: has active formal competency). ▮▮

      ✗ [Elli Citator: has active formal competency] .... ▭▭

        • formal competency=[can gain an understanding of
         unfamiliar formal structures–understands formal struc-
         tures intuitively]

      ✓ Dave Lopper: has active formal competency .... ▭▬

        • formal competency=can create formal structures independently

○ ✓ **not recommended:** MisUse Cases .... ▭▬

  ✗ [this option is neccessary] .... ▭▬

    ✗ [excluded: Lauesen Group Interview Technique] .... ▭▭

    ✗ [excluded: Prototyping] .... ▭▭

  ✗ [this option is feasible] .... ▭▭

    ▲ (Client: has passive formal competency) .... ▮▭

      ▲ (BigCorp: has passive formal competency) .... ▮▭

        • formal competency=[none–can gain
         an understanding of unfamiliar formal structures]

    ✗ [chosen: Use Cases] .... ▭▭

    ▲ (Team Member [any 1 entity]: has active formal competency). ▮▮

      ✗ [Elli Citator: has active formal competency] .... ▭▭

        • formal competency=[can gain an understanding of
         unfamiliar formal structures–understands formal struc-
         tures intuitively]

      ✓ Dave Lopper: has active formal competency .... ▭▬

        • formal competency=can create formal structures independently

○ ✓ **not recommended:** BPMN . . . . [▮]

  ✕ [this option is neccessary] . . . . [   ]

    ✕ [excluded: Lauesen Group Interview Technique] . . . . [   ]

    ✕ [chosen: Practice: Elicit Tasks and Business Process] . . . . [   ]

    ✕ [excluded: Stakeholder Analysis] . . . . [   ]

    ✕ [excluded: Prototyping] . . . . [   ]

  ▲ (this option is feasible) . . . . [███   ]

    ▲ (Client: has passive formal competency) . . . . [███   ]

      ▲ (BigCorp: has passive formal competency) . . . . [████   ]

        • formal competency=[none–can gain
an understanding of unfamiliar formal structures]

    ▲ (Team Member [any 1 entity]: has active formal competency) . [████████]

      ✕ [Elli Citator: has active formal competency] . . . . [   ]

        • formal competency=[can gain an understanding of
unfamiliar formal structures–understands formal structures intuitively]

      ✓ Dave Lopper: has active formal competency . . . . [▮]

        • formal competency=can create formal structures independently

○ ✓ **not recommended:** EPC Modelling . . . . [▮]

  ✕ [this option is neccessary] . . . . [   ]

    ✕ [excluded: Lauesen Group Interview Technique] . . . . [   ]

    ✕ [chosen: Practice: Elicit Tasks and Business Process] . . . . [   ]

    ✕ [excluded: Stakeholder Analysis] . . . . [   ]

    ✕ [excluded: Prototyping] . . . . [   ]

  ▲ (this option is feasible) . . . . [███   ]

    ▲ (Client: has passive formal competency) . . . . [███   ]

      ▲ (BigCorp: has passive formal competency) . . . . [████   ]

        • formal competency=[none–can gain
an understanding of unfamiliar formal structures]

    ▲ (Team Member [any 1 entity]: has active formal competency) . [████████]

      ✕ [Elli Citator: has active formal competency] . . . . [   ]

        • formal competency=[can gain an understanding of
unfamiliar formal structures–understands formal structures intuitively]

      ✓ Dave Lopper: has active formal competency . . . . [▮]

        • formal competency=can create formal structures independently

○ (✓) **not recommended:** Quality Function Deployment . . . . �ю▮

  (✗) [this option is neccessary] . . . . ▮□

    ✗ [excluded: Lauesen Group Interview Technique] . . . . ▮□

  (✗) [NOT Project: the duration is short] . . . . ▮□

    (✗) [Alpha: the duration is short] . . . . ▮□

      • project duration=[6.0 months–9.0 months]

(✗) [this option is feasible] . . . . ▮□

  ▲ (Client: has passive formal competency) . . . . ▮□

    ▲ (BigCorp: has passive formal competency) . . . . ▮□

      • formal competency=[none–can gain
        an understanding of unfamiliar formal structures]

  ▲ (Project: the team is small) . . . . □▮□

    ▲ (Alpha: the team is small) . . . . □▮□

      • team size=[3 persons–5 persons]

  ▲ (Team Member [any 1 entity]: has active formal competency). ▮

    ✗ [Elli Citator: has active formal competency] . . . . ▮□

      • formal competency=[can gain an understanding of
        unfamiliar formal structures–understands formal struc-
        tures intuitively]

    ✓ Dave Lopper: has active formal competency . . . . □▮

      • formal competency=can create formal structures independently

  (✗) [NOT Project: the duration is short] . . . . ▮□

    (✗) [Alpha: the duration is short] . . . . ▮□

      • project duration=[6.0 months–9.0 months]

○ (✓) **not recommended:** Goal-Domain Analysis .... ▭▮

   (✗) [this option is neccessary] .... ▮▭

      ✗ [excluded: Lauesen Group Interview Technique] .... ▯

      ✗ [excluded: Cost-Benefit Analysis] .... ▯

     (✗) [NOT Project: the duration is short] .... ▮▭

       (✗) [Alpha: the duration is short] .... ▮▭

         • project duration=[6.0 months–9.0 months]

      ✗ [excluded: Prototyping] .... ▯

    ▲ (this option is feasible) .... ▮▭

      ▲ (Client: has passive formal competency) .... ▮▭

        ▲ (BigCorp: has passive formal competency) .... ▮▭

           • formal competency=[none–can gain
            an understanding of unfamiliar formal structures]

      ▲ (Team Member [any 1 entity]: has active formal competency). ▮▮

        ✗ [Elli Citator: has active formal competency] .... ▯

           • formal competency=[can gain an understanding of
            unfamiliar formal structures–understands formal struc-
            tures intuitively]

       ✓ Dave Lopper: has active formal competency .... ▭▮

          • formal competency=can create formal structures independently

● (✓) **recommended:** Cost-Benefit Analysis .... ▭▮

   (✓) this option is neccessary .... ▭▮

      ✓ chosen: Practice: Determine Scope .... ▭▮

      ✓ excluded: Study Similar Companies .... ▭▮

      ✓ excluded: Risk Workshop .... ▭▮

     (✓) Project: the duration is short .... ▭▮

       (✓) Alpha: the duration is short .... ▭▮

         • project duration=[6.0 months–9.0 months]

    ✓ this option is feasible (after applying feasibility offset 0.2) .... ▭▮

      ✓ this option is feasible .... ▭▮

○ ✓ **not recommended:** On-Site Customer .... ▮▭

    ✗ [this option is neccessary] .... ▮▭

       ✗ [excluded: Pseudo-On-Site Customer] .... ▮▭

    ✗ [this option is feasible] .... ▮▭

       ✗ [Client: is always available on my site] .... ▮▭

          ✗ [BigCorp: is always available on my site] .... ▮▭

            • reachability=always by phone

       ✗ [excluded: Pseudo-On-Site Customer] .... ▮▭

● ✓ **recommended:** Pseudo-On-Site Customer .... ▭▮

    ✓ this option is neccessary .... ▭▮

       ✓ chosen: Practice: Integrate Stakeholders .... ▭▮

       ✓ excluded: On-Site Customer .... ▭▮

    ✓ this option is feasible (after applying feasibility offset 0.2) .... ▭▮

       ✓ this option is feasible .... ▭▮

          ✓ excluded: On-Site Customer .... ▭▮

○ ✓ **not recommended:** Document Studies .... ▭▮

    ✗ [this option is neccessary] .... ▮▭

       ✗ [chosen: Practice: Elicit Tasks and Business Process] .... ▮▭

       ✗ [excluded: Prototyping] .... ▮▭

● ✓ **recommended:** Stakeholder Analysis .... ▭▮

    ✓ this option is neccessary .... ▭▮

       ✓ excluded: Focus Groups .... ▭▮

       ✓ chosen: Practice: Identify Stakeholders and Sources .... ▭▮

       ✓ excluded: Domain Workshops .... ▭▮

       ✓ excluded: Design Workshops .... ▭▮

    ✓ this option is feasible (after applying feasibility offset 0.2) .... ▭▮

       ✓ this option is feasible .... ▭▮

● ✓ **recommended:** Lauesen Group Interview Technique . . . . ▭

   ✓ this option is neccessary . . . . ▭

      ✓ excluded: Task Demonstration . . . . ▭

      ✓ excluded: Domain Workshops . . . . ▭

      ✓ excluded: Design Workshops . . . . ▭

      ✓ excluded: Lauesen Interview Technique . . . . ▭

      ✓ excluded: User Observation . . . . ▭

      ✓ excluded: Focus Groups . . . . ▭

      ✓ excluded: Quality Function Deployment . . . . ▭

      ✓ excluded: Goal-Domain Analysis . . . . ▭

      ✓ chosen: Practice: Elicit Goals . . . . ▭

   ✓ this option is feasible (after applying feasibility offset o.2) . . . . ▭

      ✓ this option is feasible . . . . ▭

○ ✓ **not recommended:** Lauesen Interview Technique . . . . ▭

   ✗ [this option is neccessary] . . . . ▭

      ✗ [chosen: Practice: Elicit Tasks and Business Process] . . . . ▭

      ✗ [excluded: Lauesen Group Interview Technique] . . . . ▭

      ✗ [excluded: Stakeholder Analysis] . . . . ▭

      ✗ [excluded: Prototyping] . . . . ▭

○ ✓ **not recommended:** User Observation . . . . ▭

   ✗ [this option is neccessary] . . . . ▭

      ✗ [chosen: Practice: Elicit Tasks and Business Process] . . . . ▭

      ✗ [excluded: Lauesen Group Interview Technique] . . . . ▭

      ✗ [excluded: Stakeholder Analysis] . . . . ▭

      ✗ [excluded: Prototyping] . . . . ▭

   ▲ (this option is feasible) . . . . ▮

      ✓ User Group: is available by phone on appointment . . . . ▭

         ✓ UseWorks: is available by phone on appointment . . . . ▭

            • reachability=[by phone on appointment–always by phone]

      ▲ (User Group: can be visited in their surroundings) . . . . ▮

         ▲ (UseWorks: can be visited in their surroundings) . . . . ▮

○ ✓ **not recommended:** Task Demonstration..............................▮ ☐▮

   ✗ [this option is neccessary].......................................▮ ☐▮

     ✗ [excluded: Lauesen Group Interview Technique] ....▮ ☐▮

     ✗ [chosen: Practice: Elicit Tasks and Business Process].........▮ ☐▮

     ✗ [excluded: Stakeholder Analysis] ....▮ ☐▮

     ✗ [excluded: Prototyping] ....▮ ☐▮

  ▲ (this option is feasible).......................................▮▮▮

    ✓ User Group: is available by phone on appointment...........☐ ☐▮

      ✓ UseWorks: is available by phone on appointment ....☐ ☐▮

        • reachability=[by phone on appointment–always by phone]

    ▲ (User Group: can be visited in their surroundings)...........▮▮▮

      ▲ (UseWorks: can be visited in their surroundings) ....▮▮▮

○ ✓ **not recommended:** Questionnaires ....☐ ☐▮

   ✗ [this option is neccessary].......................................▮ ☐▮

  ▲ (this option is feasible).......................................▮▮▮

    ✓ chosen: Practice: Identify Stakeholders and Sources ....☐ ☐▮

    ▲ (NOT External Stakeholder: the group is small enough for workshops)..▮▮▮

      ▲ (BigCorp: the group is small enough for workshops) .....▮▮▮

      (✗) [UseWorks: the group is small enough for workshops]...▮▮ ☐.

        • size=[10 persons–15 persons]

    ▲ (Team Member [any 1 entity]: has basic knowledge of the application domain)..▮▮▮

      ▲ (Dave Lopper: has basic knowledge of the application domain) ....▮▮▮

      ▲ (Elli Citator: has basic knowledge of the application domain) ....▮▮▮

    ✓ chosen: Practice: Integrate Stakeholders ....☐ ☐▮

    ✗ [chosen: Practice: Elicit Tasks and Business Process].........▮ ☐▮

    ✓ chosen: Practice: Elicit Goals ....☐ ☐▮

○ ✓ **not recommended:** Brainstorming . . . .

    ✕ [this option is neccessary] . . . .

       ✕ [excluded: Lauesen Group Interview Technique] . . . .

       ✕ [chosen: Practice: Elicit Tasks and Business Process] . . . .

       ✕ [excluded: Stakeholder Analysis] . . . .

       ✕ [excluded: Prototyping] . . . .

    ▲ (this option is feasible) . . . .

       ▲ (Team Member [any 1 entity]: possesses good visionary skills)

          ✓ Dave Lopper: possesses good visionary skills . . . .

            • innovativeness=[good–excellent]

          ▲ (Elli Citator: possesses good visionary skills) . . . .

            • innovativeness=[bad–good]

       ▲ (External Stakeholder [any 1 entity]: possesses good visionary skills) . . . .

          ▲ (BigCorp: possesses good visionary skills) . . . .

            • innovativeness=[average–good]

          ✕ [UseWorks: possesses good visionary skills] . . . .

            • innovativeness=[unacceptable–bad]

○ ✓ **not recommended:** Focus Groups . . . .

    ✕ [this option is neccessary] . . . .

       ✕ [chosen: Practice: Elicit Tasks and Business Process] . . . .

       ✕ [excluded: Lauesen Group Interview Technique] . . . .

       ✕ [excluded: Stakeholder Analysis] . . . .

       ✕ [excluded: Prototyping] . . . .

    ✕ [this option is feasible] . . . .

       ✕ [User Group: is available for meetings] . . . .

          ✕ [UseWorks: is available for meetings] . . . .

            • reachability=[by phone on appointment–always by phone]

○ ✓ **not recommended:** Domain Workshops .... ▮▭▮

   ✗ [this option is neccessary] .... ▮▭▯

      ✗ [chosen: Practice: Elicit Tasks and Business Process] .... ▮▭▯

      ✗ [excluded: Lauesen Group Interview Technique] .... ▮▭▯

      ✗ [excluded: Cost-Benefit Analysis] .... ▮▭▯

      ✗ [excluded: Stakeholder Analysis] .... ▮▭▯

      ✗ [excluded: Prototyping] .... ▮▭▯

   ✗ [this option is feasible] .... ▮▭▯

      ✗ [User Group: is available for meetings] .... ▮▭▯

         ✗ [UseWorks: is available for meetings] .... ▮▭▯

           • reachability=[by phone on appointment–always by phone]

○ ✓ **not recommended:** Design Workshops .... ▮▭▮

   ✗ [this option is neccessary] .... ▮▭▯

      ✗ [chosen: Practice: Elicit Tasks and Business Process] .... ▮▭▯

      ✗ [excluded: Lauesen Group Interview Technique] .... ▮▭▯

      ✗ [excluded: Stakeholder Analysis] .... ▮▭▯

      ✗ [excluded: Prototyping] .... ▮▭▯

   ✗ [this option is feasible] .... ▮▭▯

      ✗ [User Group: is available for meetings] .... ▮▭▯

         ✗ [UseWorks: is available for meetings] .... ▮▭▯

           • reachability=[by phone on appointment–always by phone]

      ✗ [Project: product is a GUI application] .... ▮▭▯

         ✗ [Alpha: product is a GUI application] .... ▮▭▯

           • product type=GUI

○ ✓ **not recommended:** Features from Task Descriptions .... ▮▭▮

   ✗ [this option is neccessary] .... ▮▭▯

      ✗ [excluded: Lauesen Group Interview Technique] .... ▮▭▯

      ✗ [excluded: Prototyping] .... ▮▭▯

○ ✓ **not recommended:** Tasks & Support .... ▭
     ✗ [this option is neccessary] .... ▭
        ✗ [excluded: Lauesen Group Interview Technique] .... ▭
        ✗ [excluded: Prototyping] .... ▭
○ ✓ **not recommended:** Task-Feature Matrix .... ▭
     ✗ [this option is neccessary] .... ▭
        ✗ [excluded: Lauesen Group Interview Technique] .... ▭
        ✗ [excluded: Prototyping] .... ▭
● ✓ **recommended:** Prototyping .... ▭
     ✓ this option is neccessary .... ▭
        ✓ excluded: Tasks & Support .... ▭
        ✓ excluded: Task Demonstration .... ▭
        ✓ excluded: User Observation .... ▭
        ✓ excluded: Features from Task Descriptions .... ▭
        ✓ chosen: Practice: Elicit Non-Functional Requirements .... ▭
        ✓ excluded: Study Similar Companies .... ▭
        ✓ excluded: Goal-Domain Analysis .... ▭
        ✓ excluded: Document Studies .... ▭
        ✓ excluded: Pilot Experiments .... ▭
        ✓ excluded: Ask Suppliers to Obtain Ideas about Useful Features .... ▭
        ✓ chosen: Hypothesis: Ensure completeness of elicitation .... ▭
        ✓ excluded: Task-Feature Matrix .... ▭
     ✓ this option is feasible (after applying feasibility offset 0.2) .... ▭
        ✓ this option is feasible .... ▭
○ ✓ **not recommended:** Pilot Experiments .... ▭
     ✗ [this option is neccessary] .... ▭
        ✗ [excluded: Lauesen Group Interview Technique] .... ▭
        ✗ [Project: product is a COTS application] .... ▭
           ✗ [Alpha: product is a COTS application] .... ▭
              ● product type=GUI
        ✗ [excluded: Cost-Benefit Analysis] .... ▭
        ✗ [excluded: Prototyping] .... ▭

○ ✓ **not recommended:** Study Similar Companies . . . . □▬

    ✗ [this option is neccessary] . . . . □▬

        ✗ [excluded: Lauesen Group Interview Technique] . . . . □▬

        ✗ [excluded: Cost-Benefit Analysis] . . . . □▬

        ✗ [excluded: Prototyping] . . . . □▬

    ✗ [this option is feasible] . . . . ■□

        ✗ [NOT Project: many features have to be realised] . . . . ■□

            ✗ [Alpha: many features have to be realised] . . . . ■□

            • expected number of features=[25 Features–50 Features]

○ ✓ **not recommended:** Analysis of Project Risks . . . . □▬

    ✗ [this option is neccessary] . . . . □▬

        ✗ [excluded: Cost-Benefit Analysis] . . . . □▬

        ✗ [excluded: Prototyping] . . . . □▬

○ ✓ **not recommended:** Ask Suppliers to Obtain Ideas about Useful Features . . . . □▬

    ✗ [this option is neccessary] . . . . □▬

        ✗ [excluded: Lauesen Group Interview Technique] . . . . □▬

        ✗ [chosen: Practice: Elicit Tasks and Business Process] . . . . □▬

        ✗ [excluded: Cost-Benefit Analysis] . . . . □▬

        ✗ [excluded: Stakeholder Analysis] . . . . □▬

        ✗ [excluded: Prototyping] . . . . □▬

○ ✓ **not recommended:** Negotiation Technique: Mutual Stakeholder Analysis . . . . □▬

    ✗ [this option is neccessary] . . . . □▬

        ✗ [External Stakeholder: there are conflicts with other stakeholders] . . . . □▮

            ✗ [UseWorks: there are conflicts with other stakeholders] . . . □▮

            • conflicts between stakeholders=mostly
             false

        ✗ [chosen: Practice: Elicit Tasks and Business Process] . . . . □

○ ✓ **not recommended:** Negotiation Technique: Moderated Group Discussion with Conflict Resolution ....

    ✗ [this option is neccessary].......................................

        ✗ [External Stakeholder: there are conflicts with other stakeholders]...

            ✗ [UseWorks: there are conflicts with other stakeholders]...

                • conflicts between stakeholders=mostly false

        ✗ [chosen: Practice: Elicit Tasks and Business Process].........

○ ✓ **not recommended:** Risk Workshop ....

    ✗ [this option is neccessary].......................................

        ✗ [Project: application domain is critical] ....

            ✗ [Alpha: application domain is critical] ....

                • criticality of the application domain=[insubstantial–well below project cost]

        ✗ [chosen: Practice: Elicit Tasks and Business Process].........

        ✗ [excluded: Cost-Benefit Analysis] ....

        ✗ [excluded: Prototyping] ....

● ✗ [**recommended:** Hypothesis: Advanced elicitation practices should be carried out.]..

    • chosen by user

    ✗ [this option is neccessary].......................................

        ✗ [Project: application domain is critical] ....

            ✗ [Alpha: application domain is critical] ....

                • criticality of the application domain=[insubstantial–well below project cost]

● (✓) **recommended:** Hypothesis: Risks and consequences must be elicited ..

    (✓) this option is neccessary ....

        ✓ Project: many features have to be realised...................

            ✓ Alpha: many features have to be realised...............

                • expected number of features=[25 Features–50 Features]

        (✓) Project: the duration is short............................

            (✓) Alpha: the duration is short ....

                • project duration=[6.0 months–9.0 months]

    ✓ this option is feasible (after applying feasibility offset 0.2) ....

        ✓ this option is feasible ....

            ✓ chosen: Cost-Benefit Analysis...........................

            ✓ chosen: Prototyping ....

● ✓ **recommended:** Hypothesis: Ensure completeness of elicitation .... ▭▬

   ✓ this option is neccessary .... ▭▬

      ● always

   ✓ this option is feasible (after applying feasibility offset o.2) .... ▭▬

      ✓ this option is feasible .... ▭▬

         ✓ chosen: Prototyping .... ▭▬

C.3 Tailoring Context Ratings

C.3.1 Overview

✓ it is adequate that it is *tendentially true* that External Stakeholder: is available for discussions .... ▭▬

✓ it is adequate that it is *false* that Project: product is a COTS application .... ▭▬

✓ it is adequate that it is *ambivalent* that any 1 Stakeholder: possesses good visionary skills.. ▭▬

✓ it is adequate that it is *false* that External Stakeholder: is available for meetings .... ▭▬

(✗) [it is adequate that it is *tendentially false* that any 1 Team Member: has basic experience with requirements management]..................... ▬▭

✓ it is adequate that it is *tendentially true* that Project: the duration is short .... ▭▬

✓ it is adequate that it is *false* that External Stakeholder: there are conflicts with other stakeholders.. ▭█▬

✓ it is adequate that it is *ambivalent* that Stakeholder: has passive formal competency .... ▭▬

✓ it is adequate that it is *true* that External Stakeholder: is available by phone on appointment .... ▭▬

✓ it is adequate that it is *ambivalent* that Project: documentation relevant to the project is available .... ▭▬

✓ it is adequate that it is *false* that Project: product is a GUI application.. ▭▬

✓ it is adequate that it is *true* that External Stakeholder: the group is big enough for workshops..

✓ it is adequate that it is *ambivalent* that User Group: can be visited in their surroundings..

✓ it is adequate that it is *false* that External Stakeholder: is always available on my site..

✓ it is adequate that it is *ambivalent* that any 1 Team Member: has basic knowledge of the application domain....................................

✓ it is adequate that it is *ambivalent* that Project: the team is small.......

✓ it is adequate that it is *tendentially true* that External Stakeholder: the group is small enough for workshops..................................

✓ it is adequate that it is *ambivalent* that any 1 Stakeholder: has active formal competency..

✓ it is adequate that it is *ambivalent* that External Stakeholder: the group is small enough for workshops....................................

✓ it is adequate that it is *ambivalent* that any 1 Stakeholder: possesses good visionary skills..

(✓) it is adequate that it is *true* that Project: many features have to be realised..

✓ it is adequate that it is *false* that Project: application domain is critical.

C.3.2 Justifications

✓ it is adequate that it is *tendentially true* that External Stakeholder: is available for discussions..

   (✓) User Group: is available for discussions.........................

      (✓) UseWorks: is available for discussions......................

         • reachability=[by phone on appointment–always by phone]

    ✗ [it is necessary that the constraint holds].........................

      ✗ [chosen: Pilot Experiments]................................

   ✓ it is feasible that the constraint holds............................

✓ **it is adequate that it is** *false* **that** Project: product is a COTS application . . . .

    ✕ [Project: product is a COTS application] . . . .

        ✕ [Alpha: product is a COTS application] . . . .

            • product type=GUI

    ✕ [it is necessary that the constraint holds] . . . .

    ✕ [it is feasible that the constraint holds] . . . .

        ✕ [chosen: Pilot Experiments] . . . .

        ✕ [chosen: Practice: Elicit Tasks and Business Process] . . . .

        ✕ [excluded: Hypothesis: Advanced elicitation practices should be carried out.] . . . .

✓ **it is adequate that it is** *ambivalent* **that** any 1 Stakeholder: possesses good visionary skills . . . .

    ▲ (Team Member [any 1 entity]: possesses good visionary skills) . . . .

        ✓ Dave Lopper: possesses good visionary skills . . . .

            • innovativeness=[good–excellent]

        ▲ (Elli Citator: possesses good visionary skills) . . . .

            • innovativeness=[bad–good]

    ✕ [it is necessary that the constraint holds] . . . .

        ✕ [chosen: Brainstorming] . . . .

    ✓ it is feasible that the constraint holds . . . .

✓ **it is adequate that it is** *false* **that** External Stakeholder: is available for meetings . . . .

    ✕ [User Group: is available for meetings] . . . .

        ✕ [UseWorks: is available for meetings] . . . .

            • reachability=[by phone on appointment–always by phone]

    ✕ [it is necessary that the constraint holds] . . . .

        ✕ [chosen: Focus Groups] . . . .

        ✕ [chosen: Design Workshops] . . . .

        ✕ [chosen: Domain Workshops] . . . .

    ✓ it is feasible that the constraint holds . . . .

(✗) [it is adequate that it is *tendentially false* that any 1 Team Member: has basic experience with requirements management] . . . .

   (✗) [Team Member [any 1 entity]: has basic experience with requirements management] . . . .

      (✗) [Elli Citator: has basic experience with requirements management] . . . .

         • experience with requirements management=1.2 person years

      ✗ [Dave Lopper: has basic experience with requirements management] . . . .

         • experience with requirements management=0.5 person years

   ✓ it is necessary that the constraint holds . . . .

      ✓ chosen: Hypothesis: Advanced elicitation practices should be carried out. . . . .

✓ it is adequate that it is *tendentially true* that Project: the duration is short . . . .

   (✓) Project: the duration is short . . . .

      (✓) Alpha: the duration is short . . . .

         • project duration=[6.0 months–9.0 months]

   ✗ [it is necessary that the constraint holds] . . . .

   ✓ it is feasible that the constraint holds . . . .

      ✓ excluded: Quality Function Deployment . . . .

      ✓ chosen: Hypothesis: Risks and consequences must be elicited .

      ✓ chosen: Cost-Benefit Analysis . . . .

✓ it is adequate that it is *false* that External Stakeholder: there are conflicts with other stakeholders . . . .

   ✗ [External Stakeholder: there are conflicts with other stakeholders]

      ✗ [UseWorks: there are conflicts with other stakeholders] . . . .

         • conflicts between stakeholders=mostly false

   ✗ [it is necessary that the constraint holds] . . . .

   ✗ [it is feasible that the constraint holds] . . . .

      • never

✓ **it is adequate that it is *ambivalent* that** Stakeholder: has passive formal competency . . . .

    ▲ (Client: has passive formal competency) . . . .

       ▲ (BigCorp: has passive formal competency) . . . .

          • formal competency=[none–can gain an understanding of unfamiliar formal structures]

    × [it is necessary that the constraint holds] . . . .

       × [chosen: BPMN] . . . .

       × [chosen: Goal-Domain Analysis] . . . .

       × [chosen: Quality Function Deployment] . . . .

       × [chosen: Use Cases] . . . .

       × [chosen: EPC Modelling] . . . .

       × [chosen: MisUse Cases] . . . .

    ✓ it is feasible that the constraint holds . . . .

✓ **it is adequate that it is *true* that** External Stakeholder: is available by phone on appointment . . . .

    ✓ User Group: is available by phone on appointment . . . .

       ✓ UseWorks: is available by phone on appointment . . . .

          • reachability=[by phone on appointment–always by phone]

    ▲ (it is necessary that the constraint holds) . . . .

       ▲ (User Group: can be visited in their surroundings) . . . .

         ▲ (UseWorks: can be visited in their surroundings) . . . .

       × [chosen: User Observation] . . . .

       × [chosen: Task Demonstration] . . . .

    ✓ it is feasible that the constraint holds . . . .

✓ **it is adequate that it is *ambivalent* that** Project: documentation relevant to the project is available . . . .

    ▲ (Project: documentation relevant to the project is available) . . . .

       ▲ (Alpha: documentation relevant to the project is available) . . .

          • documentation relevant to the project is available=[mostly false–mostly true]

    × [it is necessary that the constraint holds] . . . .

       × [chosen: Document Studies] . . . .

    ✓ it is feasible that the constraint holds . . . .

✓ **it is adequate that it is *false* that** Project: product is a GUI application . . ▭

  ✗ [Project: product is a GUI application] . . . . ▭

    ✗ [Alpha: product is a GUI application] . . . . ▭

      • product type=GUI

  ✗ [it is necessary that the constraint holds] . . . . ▭

  ✗ [it is feasible that the constraint holds] . . . . ▭

    ✗ [chosen: Practice: Elicit Tasks and Business Process] . . . . ▭

    ✗ [excluded: Hypothesis: Advanced elicitation practices should
be carried out.] . . . . ▭

✓ **it is adequate that it is *true* that** External Stakeholder: the group is big
enough for workshops . . . . ▭

  ✓ User Group: the group is big enough for workshops . . . . ▭

    ✓ UseWorks: the group is big enough for workshops . . . . ▭

      • size=[10 persons–15 persons]

  ✗ [it is necessary that the constraint holds] . . . . ▭

    ✗ [chosen: Focus Groups] . . . . ▭

    ✗ [chosen: Design Workshops] . . . . ▭

    ✗ [chosen: Domain Workshops] . . . . ▭

  ✓ it is feasible that the constraint holds . . . . ▭

✓ **it is adequate that it is *ambivalent* that** User Group: can be visited in
their surroundings . . . . ▭

  ▲ (User Group: can be visited in their surroundings) . . . . ▉

    ▲ (UseWorks: can be visited in their surroundings) . . . . ▉

  ✗ [it is necessary that the constraint holds] . . . . ▭

    ✗ [chosen: User Observation] . . . . ▭

    ✗ [chosen: Task Demonstration] . . . . ▭

  ✓ it is feasible that the constraint holds . . . . ▭

✓ **it is adequate that it is *false* that** External Stakeholder: is always available on my site . . . .

  ✗ [Client: is always available on my site] . . . .

    ✗ [BigCorp: is always available on my site] . . . .

      • reachability=always by phone

  ✗ [it is necessary that the constraint holds] . . . .

    ✗ [chosen: On-Site Customer] . . . .

  ✓ it is feasible that the constraint holds . . . .

✓ **it is adequate that it is *ambivalent* that** any 1 Team Member: has basic knowledge of the application domain . . . .

  ▲ (Team Member [any 1 entity]: has basic knowledge of the application domain) . . . .

    ▲ (Dave Lopper: has basic knowledge of the application domain) . . . .

    ▲ (Elli Citator: has basic knowledge of the application domain) .

  ✗ [it is necessary that the constraint holds] . . . .

    ✗ [chosen: Questionnaires] . . . .

  ✓ it is feasible that the constraint holds . . . .

✓ **it is adequate that it is *ambivalent* that** Project: the team is small . . . .

  ▲ (Project: the team is small) . . . .

    ▲ (Alpha: the team is small) . . . .

      • team size=[3 persons–5 persons]

  ✗ [it is necessary that the constraint holds] . . . .

    ✗ [chosen: Quality Function Deployment] . . . .

  ✓ it is feasible that the constraint holds . . . .

✓ **it is adequate that it is *tendentially true* that** External Stakeholder: the group is small enough for workshops . . . .

  (✓) User Group: the group is small enough for workshops . . . .

    (✓) UseWorks: the group is small enough for workshops . . . .

      • size=[10 persons–15 persons]

  ✗ [it is necessary that the constraint holds] . . . .

    ✗ [chosen: Design Workshops] . . . .

    ✗ [chosen: Domain Workshops] . . . .

  ✓ it is feasible that the constraint holds . . . .

✓ **it is adequate that it is *ambivalent* that** any 1 Stakeholder: has active formal competency . . . .

▲ (Team Member [any 1 entity]: has active formal competency) . . . . .

✗ [Elli Citator: has active formal competency] . . . .

- formal competency=[can gain an understanding of unfamiliar formal structures–understands formal structures intuitively]

✓ Dave Lopper: has active formal competency . . . .

- formal competency=can create formal structures independently

✗ [it is necessary that the constraint holds] . . . .

✗ [chosen: BPMN] . . . .

✗ [chosen: Goal-Domain Analysis] . . . .

✗ [chosen: Quality Function Deployment] . . . .

✗ [chosen: Use Cases] . . . .

✗ [chosen: EPC Modelling] . . . .

✗ [chosen: MisUse Cases] . . . .

✓ it is feasible that the constraint holds . . . .

✓ **it is adequate that it is *ambivalent* that** External Stakeholder: the group is small enough for workshops . . . .

▲ (External Stakeholder: the group is small enough for workshops) .

▲ (BigCorp: the group is small enough for workshops) . . . .

(✓) UseWorks: the group is small enough for workshops . . . .

- size=[10 persons–15 persons]

✗ [it is necessary that the constraint holds] . . . .

✓ it is feasible that the constraint holds . . . .

✓ excluded: Questionnaires . . . .

✓ it is adequate that it is *ambivalent* that any 1 Stakeholder: possesses good visionary skills.. [bar]

  ▲ (External Stakeholder [any 1 entity]: possesses good visionary skills) .... [bar]

    ▲ (BigCorp: possesses good visionary skills).................... [bar]

      • innovativeness=[average–good]

    ✗ [UseWorks: possesses good visionary skills] .... [bar]

      • innovativeness=[unacceptable–bad]

  ✗ [it is necessary that the constraint holds] .... [bar]

    ✗ [chosen: Brainstorming] .... [bar]

  ✓ it is feasible that the constraint holds............................ [bar]

(✓) it is adequate that it is *true* that Project: many features have to be realised .... [bar]

  ✓ Project: many features have to be realised .... [bar]

    ✓ Alpha: many features have to be realised.................... [bar]

      • expected number of features=[25 Features–50 Features]

  ✗ [it is necessary that the constraint holds] .... [bar]

  (✓) it is feasible that the constraint holds............................ [bar]

    ✓ excluded: Study Similar Companies......................... [bar]

    ✓ chosen: Hypothesis: Risks and consequences must be elicited. [bar]

    (✓) Project: the duration is short.............................. [bar]

      (✓) Alpha: the duration is short .... [bar]

        • project duration=[6.0 months–9.0 months]

✓ it is adequate that it is *false* that Project: application domain is critical . [bar]

  ✗ [Project: application domain is critical]......................... [bar]

    ✗ [Alpha: application domain is critical] .... [bar]

      • criticality of the application domain=[insubstantial–well below project cost]

  ✗ [it is necessary that the constraint holds] .... [bar]

  ✓ it is feasible that the constraint holds............................ [bar]

    ✓ excluded: Practice: Elicit Tasks and Business Process .... [bar]

    ✓ chosen: Hypothesis: Advanced elicitation practices should be carried out. .... [bar]

D Abbreviations

AI Artificial Intelligence

API application programming interface

APTLY Agile Process Tailoring and probLem analYsis

CMM Capability Maturity Model

CMMI Capability Maturity Model Integration

CMU Carnegie Mellon University

FF fuzzy formula

FIL fuzzy interval logic

FPV fuzzy propositional variable

FTI fuzzy truth interval

FTV fuzzy truth value

GUI graphical user interface

HIS Herstellerinitiative Software[1]

ITIL IT Infrastructure Library

NNF negation normal form

OSSP Organization's Standard Software Process

QIP Quality Improvement Paradigm

RUP Rational Unified Process

SEI Software Engineering Institute

SIL Safety Integrity Level

SME small or medium enterprise

SPiCE Software Process Improvement and Capability dEtermination

---

[1]Manufacturer's Software Initiative

D Abbreviations

SW-CMM Software Capability Maturity Model

TAME Tailoring A Measured Environment

TSS tailoring support system

XP eXtreme Programming

Softwaregestütztes Anpassen von Prozessbeschreibungen

Zusammenfassung

Agile Softwareentwicklung schmälert nicht die Notwendigkeit, bewusst über Prozesse nachzudenken. Im Gegenteil müssen agile Paradigmen in die breitere Sichtweise der *angemessenen* Prozesse eingegliedert werden: Das Maß an Flexibilität, die ein Prozess zulässt, sollte immer den Erfordernissen des aktuellen Projekts angepasst werden. *Prozesstailoring* ist ein wirksames Mittel, um diese Anpassungen vorzunehmen. Allerdings ist das Anpassen von Prozessen eine schwierige Aufgabenstellung, die sowohl Erfahrung als auch gute Kenntnisse über das aktuelle Projekt erfordert. Um das Prozesstailoring zu erleichtern, und um seine weitere Verbreitung zu befördern, legen wir als Grundlage für ein Softwaresystem zur Tailoringunterstützung ein formales Tailoringrahmenwerk vor.

Unser Tailoringrahmenwerk gestattet, unter Berücksichtigung einer dreigeteilten Struktur formale *Tailoringrichtlinien* auszudrücken. Die Struktur umfasst verfügbare *Tailoring-Optionen*, ein *Tailoring-Universum*, das Metriken definiert über den Kontext, in dem ein Tailoring durchgeführt wird, und *Tailoringhypothesen*, die in der Form logischer Propositionen die Abhängigkeiten und Einschränkungen ausdrücken, denen die Auswahl einzelner Optionen mit Rücksicht auf den Kontext unterliegt. Tailoring-Optionen entsprechen binären Wahlmöglichkeiten über Bestandteile einer Prozessbeschreibung; der Tailoringvorgang besteht darin, für jeden Bestandteil zu entscheiden, ob er in der Prozessbeschreibung beibehalten werden soll, oder ob er ausgeklammert werden soll. Autoren von Tailoringrichtlinien können *Tailoringregeln* unabhängig von einzelnen Tailoring-Optionen formulieren; ein spezieller Mechanismus wandelt sie automatisch in äquivalente Tailoringhypothesen um, die sich auf jeweils eine einzelne Option beziehen. Wenn ein Prozessmodell vorliegt, dass um solche Tailoringrichtlinien ergänzt ist, kann ein auf unserem Rahmenwerk basierendes Tailoringsystem eingesetzt werden, um eine konkrete Prozessbeschreibung auf die Anforderungen eines bestimmten Kontexts zurechtzuschneidern.

Tailoring erfolgt in drei Schritten. Zuerst beschreibt der Benutzer des Tailoringsystems den aktuellen *Tailoringkontext*, indem er Messungen und Schätzungen für die im Tailoring-Universum aufgestellten Metriken vorgibt. Dann ruft er den Optimierungsalgorithmus des Systems auf, um die optimale *Tailoringkonfiguration* zu erhalten. Auf der Basis von unscharfer Logik stuft das System die gegeneinander antretenden Tailoringkonfigurationen anhand von Valuationen der Tailoringhypothesen ein, und bezieht auf diese Weise sowohl den Tailoringkontext als auch die Tailoringentscheidungen mit ein. Als letztes überprüft der Benutzer die vom System empfohlene Tailoringkonfiguration und stimmt ihr entweder zu, oder er revidiert einige der Tailoringentscheidungen und lässt das System die verbleibenden Entscheidungen erneut optimieren. Anhand eines Begründungsmechanismus kann das Tailoringsystem für jede Tailoringentscheidung offen legen, welche Gesichtspunkte des Tailoringkontexts und welche anderen Tailoringentscheidungen für oder gegen sie sprechen.

Wir schließen mit einer beispielhaften Anwendung unseres Rahmenwerks im Zusammenhang des *ReqMan*-Projekts, und besprechen verwandte Ansätze zur Tailoringunterstützung.

Inhaltsverzeichnis

Inhaltsverzeichnis

Einführung

Angemessene Prozesse durch Prozessanpassung

Das Ziel des *Software Engineering* ist – wie in jeder anderen Ingenieursdisziplin – Produkte in hoher Qualität unter Aufwendung möglichst weniger Ressourcen herzustellen. Typische Qualitätsanforderungen des Software Engineering sind Zuverlässigkeit, Effizienz, Wartbarkeit, Bedienerfreundlichkeit, und viele andere. In den späten 60er Jahren wurden sich Softwareingenieure darüber klar, dass solche Qualitätsanforderungen nur gewährleistet werden können, wenn man "einem disziplinierten Ablauf von Tätigkeiten folgt" [CG98, übertragen aus dem Englischen]. Diese Einsicht führte zur Definition von *Software-Lebenszyklen* [Roy70], und findet sich heute weitestgehend unter dem Oberbegriff *Software-Prozesse* wieder.

Frühe Softwareprozesse bezogen ihre Anregungen aus Paradigmen der klassischen Industrieproduktion. Im Laufe der Jahre stellte sich heraus, dass diese Prozesse einige Qualitätsprobleme nicht angemessen lösen konnten, die ihre Ursache in der besonderen, immateriellen Beschaffenheit von Software hatten. Diese Einsicht führte, angestoßen von der *Agilen Bewegung*, um die Jahrtausendwende zu einem radikalen Paradigmenwechsel [AgA]. Der Begriff *Agilität* wurde bald gleichbedeutend mit hochangepassten, schlanken und flexiblen Prozessen, die sich grundlegend von ihren starren, formalen und streng plangetriebenen Vorläufern unterschieden.

Mittlerweile hat die Agile Bewegung ihre rebellischen Zeiten längst beendet; agiles Gedankengut hat sich stetig im Tagesgeschäft der Softwareentwicklung verbreitet, vor allem in kleinen oder mittelgroßen Teams [Cop01, Miš05]. Zweifellos hat die agile Softwareentwicklung sowohl auf der persönlichen wie auf der unternehmerischen Ebene viele Fortschritte in Sachen Flexibilität und Individualität gebracht. Dennoch wäre es – wie bei allen Innovationen – ein Fehler, agile Methoden für einen allgemeingültigen Ersatz für traditionelle Methoden des Software-Engineerings zu halten [SR03]. Wie wir nun darlegen werden, sind Softwareprojekte dann am erfolgreichsten, wenn sie an keiner der beiden extremen Sichtweisen auf Prozesse festhalten, sondern stattdessen Aspekte sowohl starrer als auch agiler Herangehensweisen vereinen und sie mit Rücksicht auf die spezifischen Eigenschaften des Projekts gegeneinander austarieren.

Wir kommen um Prozesse nicht herum

Die erste der vier zentralen Thesen des von der *Agile Alliance* aufgestellten Agilen Manifests lautet:

> "[Wir messen] Individuen und Interaktionen [mehr Bedeutung bei] als Prozessen und Werkzeugen." [AgM, übertragen aus dem Englischen]

Diese Aussage hat zu Missverständnissen, oder zumindest zu einer eingeschränkten Sichtweise über den Prozessbegriff geführt. Sie legt nahe, dass Prozesse ein Bestandteil der Softwareentwicklung sind, der gegen andere Bestandteile abgewägt werden kann, wie etwa "Individuen" und "Interaktionen." Die irrtümliche Schlussfolgerung ist, dass Prozesse wahlfreier Bestandteil eines Softwareentwicklungsprojekts sind, die in einem größeren oder kleineren Maß einbezogen werden können [Gla01].

Das Online-Lexikon *Wikipedia* definiert einen Prozess als

> "eine natürlich auftretende oder konstruierte Abfolge von Handlungen oder Ereignissen, die möglicherweise Zeit, Platz, Kenntnisse oder andere Ressourcen beansprucht, und ein bestimmtes Endergebnis hat." [Wikd, übertragen aus dem Englischen]

Hier stoßen wir auf eine andere Sichtweise: Prozesse sind ein inhärenter Bestandteil aller Arten ergebnisorientierter Handlungen oder Ereignisse, Software-Engineering-Prozesse eingeschlossen. In Bezug auf die Softwareentwicklung, einschliesslich der agilen Softwareentwicklung, geht es folglich nicht darum, ob *weniger* oder *mehr* Prozess in einem Projekt zugelassen werden sollte, sondern darum, ob man sich der einem Projekt zugrunde liegenden Prozesse *bewusst* ist. Über Prozesse nicht nachzudenken ist eine verpasste Chance für Verbesserungen. Über einen eindeutigen Prozess zu verfügen ist auch eine der zentralen Anforderungen, die von Standards für die Software-Prozess-Bewertung und -Verbesserung aufgestellt werden; CMMI verlangt auf Ebene 3 *etablierte* Prozesse [SEI02a], und gleichermaßen erfordert SPiCE auf Ebene 3 einen *definierten* Prozess [ISO98b].[a]

Es ist also nicht überraschend, dass auch die Agile Bewegung mehrere Vorschläge für Softwareentwicklungsprozesse hervorgebracht hat, an erster Stelle *eXtreme Programming* (XP) [Bec01], aber auch andere Ansätze wie *Crystal* [Coc02] oder *Scrum* [SB02].

Angemessene Prozesse

Agilität wird für gewöhnlich mit Vorgehensweisen in Verbindung gebracht, die mehr Flexibilität bei weniger unnötigem Überbau in der alltäglichen Projektarbeit gestatten. Bislang lag der Fokus der agilen Welt hauptsächlich auf Techniken zur Erleichterung der eigentlichen Softwareerstellung. Die meisten Vertreter agiler Prozesse bieten bislang jedoch nur

---

[a]Wir behandeln CMMI und SPiCE ausführlicher in Kapitel 5.

Beschreibungen monolithischer Prozesse an, die keine Anpassung an die besonderen Erfordernisse einzelner Software-Engineering-Projekte gestatten [SR03].

Eines der grundlegenden Konzepte der Agilen Bewegung ist die Vorstellung, dass Software ein lebendiger Organismus ist, der sich genauso leicht ändern können sollte, wie es für die Anforderungen an die Software der Fall ist. Kurze Freigabezyklen bezwecken, dass Diskrepanzen zwischen der Funktionalität der Software und der tatsächlich an sie gestellten Anforderungen so früh wie möglich erkannt werden. *Refactoring*, ein agiler Ansatz zur Umstrukturierung von Quelltext nach bewährten Schemata und Prinzipien, hat sich zu einer eigenständigen Disziplin entwickelt [Fow99]. Um einen hohen Grad an Agilität sicherzustellen, sollte dieselbe Flexibilität auch für den Entwicklungsprozess gelten, der das Projekt vorantreibt. Die agile Sichtweise sollte sich nicht auf die Gestaltung von Arbeitsmethoden (das *Wie*) beschränken, sondern sollte auch die *Auswahl* der Methoden und Mittel (das *Was*) betreffen, und somit das "Gewicht" der Methoden flexibel den veränderlichen Erfordernissen des Projekts anpassen.

Das *Appropriate Process Movement* [APM] hat für Prozesse, die einer gegebenen Situation ideal angepasst sind, den Begriff *angemessener Prozess* geprägt. Anstatt schwergewichtige, strenge Prozesse und leichtgewichtige, agile Prozesse als zwei unvereinbare Gegensätze zu betrachten, sieht die Bewegung diese zwei Konzepte als die beiden Enden einer Skala, auf der jeder Prozess abhängig von der Situation unterschiedlich angeordnet werden muss:

> "Ein Prozess sollte so agil wie möglich, und so robust wie nötig sein." [Old03, übertragen aus dem Englischen]

Das Ziel ist also, sich von Prozessen in "Einheitsmaßen" zu lösen, und sich flexibleren Varianten von Prozessen zuzuwenden, die den Erfordernissen des aktuellen Projekts angepasst werden können. Diese Art der Anpassung ist als *(Prozess-)Tailoring* bekannt, vor allem im Umfeld traditioneller, schwergewichtiger und hochformalisierter Prozessmodelle wie der Rational Unified Process, oder bei Prozessreferenzmodellen wie CMMI oder SPiCE. Nichtsdestotrotz sollte Tailoring ebenso hilfreich sein in Umfeldern, die nicht durchgängig – wenn überhaupt – strenge formale Anforderungen an Projekte stellen. Tailoring kann hier genauso dazu beitragen, das ideale Gewicht von Entwicklungsprozessen einzustellen – in Einklang mit dem agilen Prinzip "so wenig wie möglich, so viel als nötig."

Tailoring

Tailoring ist eine Konfigurationsaufgabe. So, wie ein Schneider (engl. *tailor*) einen Anzug auf seinen Träger anpasst, bedeutet das Zurechtschneidern eines Prozesses, ihn passend für das aktuelle Projekt zu machen. Kein Schneider denkt sich ein neues Schnittmuster aus für jeden Anzug, den er anfertigt, stattdessen greift er auf bestehende Schnittmuster zurück, die er aufgrund einiger standardisierter Maße seines Kunden anpasst. Die Vorgehensweise beim Prozesstailoring ist ähnlich; hier besteht Tailoring darin, ein vorgegebenes, allgemeines Prozessmodell für einen vorliegenden Zweck und Zusammenhang anzupassen, hauptsächlich, indem Bestandteile der Prozessbeschreibung ausgewählt und angepasst werden [HV03]. Allerdings fanden wir unter den bestehenden Prozesstailoringansätzen nur wenige, die vorausgehende Messungen oder Abschätzungen einbeziehen – und wenn doch, dann nur in sehr begrenztem Ausmaß.[b]

Tailoring ist in Hinblick auf die Komplexität vieler Standardprozessmodelle eine schwierige Aufgabe. Die Aufgabenstellung des Tailoring wird noch überwältigender, wenn keine ausreichende Erfahrung mit einem komplexen Prozessmodell vorliegt, und keine Hilfestellung verfügbar ist, die dies ausgleichen könnte. Oftmals wird Tailoring unter solchen Umständen durch den Zugang zu verallgemeinertem Tailoringwissen erleichtert [Xu05]. Tailoring wird darüber hinaus vereinfacht, wenn Metriken verfügbar sind, anhand derer die vorliegenden Projektziele und -eigenschaften beurteilt werden können [LR93].

Thesen

Bei unserer Erhebung über bestehende Ansätze des Prozesstailorings stießen wir auf keinen Ansatz, der eine vereinheitliche, allgemeine Lösung für das Tailoringproblem liefert: Gemäß unserer obigen Beobachtungen sollte eine solche Lösung effektive Unterstützung für das Tailoring angemessener Prozesse bieten, auf der Grundlage sowohl von Erfahrungswissen, als auch von Eigenschaften des aktuellen Projekts. In dieser Arbeit wollen wir einen Beitrag leisten, diese Lücke zu schließen. Indem wir sowohl die formalen Grundlagen als auch eine konkrete Softwareimplementierung eines allgemeinen Tailoringrahmenwerks liefern, wollen wir die Voraussetzungen für softwaregestütztes Tailoring schaffen. Um zu konkretisieren, welche Erwartungen an einen effektiven Beitrag zu softwaregestütztem Tailoring gestellt werden können, stellen wir sechs Thesen auf.

Das vorrangige Ziel, und der angestrebte Beitrag unserer Arbeit ist, Prozesstailoring mithilfe einer geeigneten Softwareanwendung zu unterstützen. Unsere erste These besagt, dass dieses Ziel erreicht werden kann:

These 1 (effektive Werkzeugunterstützung für das Prozesstailoring) *Ein Software-Werkzeug kann effektive Unterstützung für das Prozesstailoring leisten.*

Die Komplexität des Tailorings beruht auf den vielen Bedingungen und Abhängigkeiten, die dabei beachtet werden müssen. Um Unterstützung für den Tailoringvorgang zu bieten, ist es erforderlich, diese Bedingungen und Abhängigkeiten in der Gestalt von Tailoringrichtlinien zu konkretisieren, die verallgemeinertes Wissen über Tailoring darstellen

---

[b]Für einen Überblick über gegenwärtige Tailoringansätze siehe Kapitel 5.

und dazu dienen, unter Beachtung der aktuellen Situation Tailoringempfehlungen zu geben. Wenn solche Richtlinien anhand eines Software-Werkzeugs angewandt werden sollen (These 1), müssen sie formal in einer geeigneten Sprache ausgedrückt werden:

These 2 (angemessene und eindeutige formale Sprache für Tailoringrichtlinien)
Tailoringrichtlinien können angemessen und eindeutig in einer formalen Sprache ausgedrückt werden.

Es gibt viele unterschiedliche Auffassungen von formaler Sprache in Gebieten wie zum Beispiel der Mathematik oder der Linguistik [Wikb]. Die Kerneigenschaft formaler Sprachen, mit der wir uns in These 2 befassen, ist die Eindeutigkeit, so dass ein Software-Werkzeug Ausdrücke in dieser Sprache interpretieren kann. Unter Angemessenheit verstehen wir, dass die Ausdruckskraft der formalen Sprache gut auf den Anwendungsbereich der Tailoringrichtlinien angepasst ist.

Um effektive Tailoringunterstützung zu bieten, beanspruchen wir weiterhin, dass ein Tailoringsystem so konstruiert werden kann, dass es konkurrierende Tailoringkonfigurationen gegeneinander abwägen kann (These 3), um die optimale Tailoringkonfiguration in Bezug auf die Tailoringrichtlinien zu finden (These 4).

These 3 (Tailoringkonfigurationen nach Wertung anordnen) *Formale Tailoringrichtlinien können dazu verwendet werden, eine Wertung für jede mögliche Tailoringkonfiguration zu berechnen, die eine totale Ordnungsrelation über Tailoringkonfigurationen induziert und die es gestattet, sie unter dem Aspekt ihrer Eignung in einem gegebenen Kontext gegeneinander abzuwägen.*

These 4 (effiziente Ermittlung der optimalen Tailoringkonfiguration) *Es ist möglich, einen effizienten Algorithmus zu erstellen, der die in Bezug auf ihre Wertung optimale Tailoringkonfiguration findet.*

Der Erfolg eines Softwarewerkzeugs für das Tailoring hängt von seiner Akzeptanz ab, die wiederum einfache Handhabung und Pflege erfordert. Das muss sowohl in Bezug auf Autoren von Tailoringrichtlinien zutreffen, als auch in Bezug auf Anwender, die die Tailoringrichtlinien über die Tailoringsoftware anwenden. Für Autoren von Tailoringrichtlinien muss das Tailoringsystem wartbar sein:

These 5 (Wartbarkeit und Skalierbarkeit) *Es ist möglich, ein Tailoringsystem zu erstellen, das die Wartbarkeit von Tailoringrichtlinien gewährleistet, selbst wenn sie umfangreich und komplex werden.*

Anwender, die sich beim Tailoring unterstützen lassen, müssen nachvollziehen können, aus welchen Gründen die Software ihre Tailoringvorschläge macht:

These 6 (transparente Bewertungen von Tailoringkonfigurationen) *Es ist möglich, eine Tailoringkonfiguration gemeinsam mit deren Bewertung so darzustellen, dass die Gründe für ihre Bewertung ohne Kenntnis der zugrunde liegenden Tailoringregeln nachvollziehbar sind.*

Aufbau dieser Arbeit

In dieser Arbeit stellen wir ein formales Rahmenwerk auf, mit dem Tailoringrichtlinien ausgedrückt und angewandt werden können. Wir beabsichtigen dabei, dieses Rahmenwerk als eine Softwarebibliothek zu implementieren, um damit die Grundlage für ein System zur Tailoringunterstützung mit grafischer Benutzeroberfläche zu schaffen.

In Kapitel 2 schaffen wir die Voraussetzungen für ein solches Tailoringrahmenwerk und liefern formale Konzepte für Optionen, die fürs Tailoring zur Verfügung stehen, für Eigenschaften des Kontexts, in dem getailort wird, und für Bedingungen, die der Beurteilung von Tailoringentscheidungen dienen auf der Grundlage sowohl des gegebenen Kontexts, als auch der Abhängigkeiten zu anderen Tailoringentscheidungen.

Aufbauend auf diesen Konzepten führen wir außerdem ein Bewertungsschema ein, das es ermöglicht, unterschiedliche Tailoringkonfigurationen gegeneinander abzuwägen. Wir stellen dann einen Optimierungsalgorithmus bereit, mit dem die optimale Tailoringkonfiguration in Bezug auf die Tailoringrichtlinien gefunden werden kann.

Ein Tailoringsystem wird zwei Hauptarten von Anwendern haben – den Prozessmodellierer, der die Tailoringrichtlinien aufstellt, und den Prozessanpasser, der das Tailoringsystem einsetzt, um ihm beim Anpassen eines konkreten Prozesses zu helfen. In Kapitel 3 schlagen wir Erweiterungen unseres Rahmenwerks vor, die beiden Gruppen die Erledigung ihrer Aufgaben erleichtern: Wie werden erläutern, wie dem Prozessmodellierer ermöglicht werden kann, Tailoringregeln flexibler zu formulieren, als es das Tailoring-Kernrahmenwerk gestattet. Um dem Prozessanpasser zu helfen, vom Tailoringsystem aufgestellte Tailoringvorschläge und -bewertungen zu verstehen, entwickeln wir außerdem einen Erklärungsmechanismus, der alle relevanten Fakten offen legt, die zu einer bestimmten Bewertung beigetragen haben.

Um die Anwendbarkeit unseres Ansatzes zu überprüfen, und um einen Eindruck seiner Arbeitsweise zu vermitteln, erläutern wir eine praktische Anwendung der Pilotimplementierung unseres Rahmenwerks in Kapitel 4. Das Beispiel ist einem Prozessrahmenwerk für das Anforderungsmanagement entnommen, das im Rahmen des Projekts *ReqMan* entwickelt wurde.

In Kapitel 5 geben wir einen Überblick über verwandte Ansätze zur Tailoringunterstützung. Wir untersuchen das V-Modell XT besonders ausführlich, weil es gegenwärtig der einzige etablierte Prozessstandard ist, der Ansätze zu softwarebasierter Tailoringunterstützung aufweist.

Wir schließen mit einer Bewertung der Ergebnisse unserer Arbeit in Kapitel 6, und weisen auf einige verbleibende Herausforderungen hin, deren zukünftige Erforschung wir für lohnend halten.

Endergebnis

Zusammenfassung

Angemessene Prozesse sind Prozesse, die auf ein vorgegebenes Umfeld so angepasst sind, dass sie so agil wie möglich, aber so rigide wie nötig sind. Das Anpassen einer Prozessbeschreibung auf ein vorgegebenes Umfeld wird Prozesstailoring genannt. Unter den von uns untersuchten bestehenden Ansätzen zum Prozesstailoring existiert keine allgemeine Lösung, die die Bereitstellung von Tailoringrichtlinien gestattet, um auf der Basis sowohl von Erfahrungswissen als auch von Eigenschaften des aktuellen Projekts durch Tailoring zu angemessenen Prozessen zu kommen. Das Ziel unserer Forschung war, diese Lücke zu schließen und die Voraussetzungen für ein softwarebasiertes System zur Tailoringunterstützung zu schaffen, das in solcher Weise Hilfe beim Prozesstailoring leistet. Folglich haben wir ein Rahmenwerk für das Tailoring aufgestellt, haben es als *Java*-Klassenbibliothek implementiert, und haben es dann im Zusammenhang eines Prozessmodells für das Anforderungsmanagement evaluiert.

Abbildung A zeigt die grundlegende Struktur unseres Tailoringrahmenwerks. Es unterscheidet zwei Ebenen, eine für abstrakte Prozessmodelle und eine für konkrete Prozessbeschreibungen. Im Verlauf des Tailoring wird eine konkrete Prozessbeschreibung von einem abstrakten Prozessmodell abgeleitet. Beide Ebenen des Rahmenwerks sind in drei Aspekte aufgegliedert. Jeder Aspekt auf der abstrakten Ebene hat eine Entsprechung auf der konkreten Ebene. Der Rest dieser Zusammenfassung folgt dieser Struktur, beginnend mit den drei Aspekten der abstrakten Ebene.

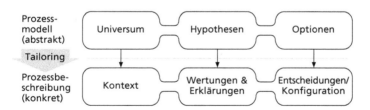

Abbildung A: Schematische Übersicht unseres Tailoringrahmenwerks

Auf der abstrakten Prozessmodell-Ebene umfasst unser Rahmenwerk drei Grundbestandteile von Tailoringrichtlinien, die alle vom Autor der Tailoringrichtlinien vorgegeben werden müssen: Ein *Tailoring-Universum* bestimmt die Typen von Entitäten, die in Tailoringsituationen relevant sind, wie z. B. *Projekt* oder *Kunde*, und die Eigenschaften jedes Typs, so wie *Dauer* oder *Erfahrung*. *Tailoring-Optionen* stehen für die binären *Ja/Nein*-Entscheidungen, die im Verlauf des Tailorings getroffen werden müssen – jede Tailoring-Option kann entweder in eine Tailoringkonfiguration aufgenommen oder aus ihr ausgeschlossen werden. Für jede Tailoring-Option existieren zwei *Tailoringhypothesen*: Eine *Erforderlichkeits-Hypothese*, die bestimmt, wann die Auswahl einer Option erforderlich ist, und eine *Zulässigkeits-Hypothese*, die bestimmt, wann die Auswahl einer Option möglich ist. Diese Tailoringhypothesen haben die Form von logischen Propositionen. Für jede Tailoring-Option formulieren sie die Abhängigkeiten zu Eigenschaften von Entitäten und zu Entscheidungen über andere Tailoring-Optionen.

Wenn der Benutzer eines Tailoringsystems ein Prozessmodell in eine konkrete, angepasste Prozessbeschreibung überführt, bestimmt er einen *Tailoringkontext* hinsichtlich der im Tailoring-Universum vorgegebenen Entitätstypen und deren Eigenschaften. Er kann Messungen oder Schätzungen von Eigenschaften als Intervalle angeben, wenn ein exakter Wert nicht bekannt ist, oder kann sogar beschließen, nicht für alle Eigenschaften Werte anzugeben oder sogar einige Entitätstypen nicht zu instanziieren. Dann fordert er das Tailoringsystem auf, die optimale Tailoringkonfiguration, d. h. die Konfiguration mit der besten Bewertung, zu finden. Das Tailoringsystem berechnet die Gesamtbewertung einer Tailoringkonfiguration aus den Bewertungen der einzelnen *Tailoringentscheidungen* über die Tailoring-Optionen, die es wiederum aus den Valuationen in der *Logik unscharfer Intervalle* ihrer jeweiligen Hypothesen ermittelt. Jede Bewertung wird als Intervall auf einer Güteskala im Bereich von 0% bis 100% dargestellt, wobei die Extreme der Bewertungsintervalle die pessimistische und die optimistische Bewertung darstellen. Die Bewertung der Tailoringkonfiguration wird aus den optimistischen Bewertungen aller Tailoringentscheidungen ermittelt. Um die Tailoringkonfiguration mit der besten Bewertung zu finden, verwendet das Tailoringsystem den T*-Optimierungsalgorithmus, den wir vom allgemein bekannten A*-Algorithmus abgeleitet haben.

Tailoringhypothesen direkt vorzugeben ist eine schwierige Aufgabe, weil sich jede Hypothese ausschließlich auf eine einzelne Tailoring-Option bezieht. Um gegenseitige Abhängigkeiten zwischen zwei oder mehreren Tailoring-Optionen auszudrücken, müssen die Hypothesen über *alle* diese Optionen geändert werden. Um die Festlegung von Tailoringhypothesen zu vereinfachen, haben wir deshalb das Konzept *allgemeiner Tailoringregeln* eingeführt, die keine Bedingungen über bestimmte Tailoring-Optionen ausdrücken, sondern stattdessen allgemeine Einschränkungen über Tailoringkonfigurationen formulieren. Wir haben die deontische Logik, ein Mitglied der Familie der Modallogiken, eingeführt, um eine solide Methode herzuleiten, ein System allgemeiner Tailoringregeln in ein äquivalentes System von Tailoringhypothesen zu überführen. Diese Transformationsmethode hat zudem den zusätzlichen Vorteil, dass sie Hypothesen über *unscharfe propositionale*

Variablen liefert, die es ermöglichen, die Übereinstimmung von Eigenschaften des Tailoringkontexts mit einer gegebenen Tailoringkonfiguration zu bewerten. Da Werte für Eigenschaften des Tailoringkontexts nur von Hand vom Benutzer bereitgestellt werden und nie vom System vorgeschlagen werden, haben die Bewertungen unscharfer propositionaler Variablen keinen Einfluss auf das Verhalten des Tailoringsystems, können aber helfen, Schwächen und Widersprüche in der Charakterisierung der Tailoringsituation durch den Benutzer festzustellen.

Um Transparenz über die vom Tailoringsystem berechneten Bewertungen zu verschaffen, und um das Vertrauen des Benutzers in die Bewertungen zu stärken, haben wir Algorithmen entworfen, die dem Tailoringsystem gestatten, Begründungen für alle Bewertungen zu liefern. Diese Begründungen haben eine einfache hierarchische Struktur von Aussagen, wobei untergeordnete Aussagen Gegebenheiten ausdrücken, die die ihnen übergeordnete Aussage stützen. Wir haben auch einen Vorschlag gemacht, auf welche Weise Begründungshierarchien dargestellt werden können, so dass sie auf verständliche Weise in einer grafischen Benutzeroberfläche angezeigt werden können.

Um ein Beispiel für die Anwendung unseres Tailoring-Rahmenwerks zu geben, haben wir aus den im *ReqMan*-Projekt aufgestellten Praktiken und Techniken des Anforderungsmanagements ein durch Tailoring anpassbares Prozessmodell aufgestellt, und haben eine von unserer Implementierung des Tailoringsystems berechnete optimale Tailoringkonfiguration dokumentiert und besprochen.

In einem Überblick über verwandte Ansätze zur Tailoringunterstützung haben wir dargelegt, dass regelbasierte Ansätze wie der unsrige viele Vorteile gegenüber fallbasierten Ansätzen aufweisen, und haben einen detaillierten Vergleich mit dem *V-Modell XT* vorgelegt, einem vor kurzem aufgestellten Prozessmodell, das unter den von uns untersuchten Prozessmodellen das erste ist, das ein Softwarewerkzeug zur Tailoringunterstützung beinhaltet.

Validierung unserer Thesen

Nachdem wir die Ergebnisse unserer Arbeit kurz zusammengefasst haben, kommen wir nun auf unsere in Abschnitt 1.2 aufgestellten Thesen zurück, um ihre Erfüllung zu überprüfen.

These 1: Effektive Werkzeugunterstützung für das Prozesstailoring

Wir haben unser Tailoring-Rahmenwerk als *Java*-Klassenbibliothek realisiert, die alle in den vorangegangenen Kapiteln aufgestellten Datenstrukturen und Algorithmen implementiert. Wir haben sie erfolgreich zur Berechnung des Fallbeispiels eingesetzt, das wir in Abschnitt 4.3 besprochen und in den Anhängen B und C vorgelegt haben. Wir wollen

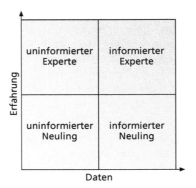

Abbildung B: Die Anforderungen an Tailoringunterstützung hängen vom Kurzfristwissen (Daten) und vom Langfristwissen (Erfahrung) des Anwenders ab.

jetzt begründen, dass ein Tailoringsystem auf der Grundlage dieser Bibliothek tatsächlich angemessene Unterstützung für das Prozesstailoring leisten kann, wie von These 1 behauptet.

Bei unserem Ansatz fürs Prozesstailoring gründen Tailoringentscheidungen auf zwei Wissensquellen – Kurzfristwissen über Kontext und Beschaffenheit des aktuellen Projekts *(Daten)*, und allgemeines Langfristwissen über Prozesse *(Erfahrung)*, das dabei hilft, die Angemessenheit verschiedener Praktiken und Werkzeuge für das aktuelle Projekt einzuschätzen. In unserem Tailoringrahmenwerk (Abschnitt 2.2) wird Kurzfristwissen als Tailoringkontext in Form von Messungen und Schätzungen für Metriken ausgedrückt, während das Langfristwissen als Menge von Tailoringhypothesen formuliert wird.

Auf der Grundlage dieser Unterscheidung können wir vier grundsätzliche Typen von Tailoring-Entscheidern feststellen (Abbildung B), die jeweils unterschiedliche Defizite bei aktuellen Daten, Erfahrung oder beidem aufweisen. Wir betrachten nun die möglichen Beiträge, die ein auf unserem Rahmenwerk aufbauendes Tailoringsystems für jede der oben genannten vier Arten von Entscheidern leisten kann.

Der informierte Experte: Konzentration auf relevante Entscheidungen

Der Tailoringvorgang beinhaltet normalerweise, eine Liste von 50 bis 200 Tailoring-Optionen durchzugehen, z. B. beim RUP oder dem *V-Modell XT*, oder bei Standards zur Prozessbewertung und -verbesserung wie SPiCE oder CMMI. In den meisten Fällen ist die Mehrzahl dieser Entscheidungen trivial und könnte auf einfache Weise vom Projektkontext abgeleitet werden, z. B. sollte ein Projekt mit hoher Sicherheitsstufe offensichtlich standardisierte Vorgehensweisen zur Gewährleistung entsprechender Sicherheitsmaßnahmen beinhalten. Während der erfahrene Projektleiter nicht auf Unterstützung zur Entscheidungsfindung angewiesen ist, kann der Tailoringvorgang nennenswert rationalisiert werden, wenn das zugrunde liegende Tailoringsystem alle hinreichend offensichtlichen Entscheidungen vorwegnimmt. Der Projektleiter erhält eine Übersicht dieser vorweggenommenen Entscheidungen und kann, nachdem er sie überprüft hat, die verbleibenden, wesentlicheren Tailoringentscheidungen treffen.

Der informierte Neuling: Überblick gewinnen

Ein unerfahrener Projektleiter mag mit allen erforderlichen Daten konfrontiert sein, kann sie vieleicht aber nicht vollständig und wirksam deuten. Ein Tailoringsystem kann mithilfe der im Tailoring-Universum definierten Entitätstypen und Eigenschaften dabei helfen, relevante Fakten und Schätzungen über das Projekt aus den vorhandenen Daten herauszuarbeiten. Das Tailoringsystem kann dann aufgrund dieser Informationen Tailoringentscheidungen vorschlagen. Weiterhin kann es Tailoringentscheidungen des Benutzers auf Konsistenz überprüfen und kann Überschneidungen und Lücken in der durch Tailoring anzupassenden Prozessbeschreibung aufdecken.

Der uninformierte Experte: unvollständige Informationen handhaben

Durch die Verwaltung von Metriken, die Schlüsselaspekte des aktuellen Projekts beschreiben, kann ein auf unserem Rahmenwerk basierendes Tailoringsystem einen strukturierten Überblick über den Projektkontext verschaffen. Das Tailoringsystem erleichtert es so, die relevantesten Informationslücken festzustellen, also jene Aspekte, die noch nicht ausreichend betrachtet wurden, aber wesentlich zu sachkundigen Tailoringentscheidungen beitragen würden. Indem es analysiert, welche Eigenschaften des Tailoringkontexts in den Hypothesen welcher Tailoring-Optionen vorkommen, kann das Tailoringsystem wesentlich zur Priorisierung der verbleibenden Informationslücken beitragen.

Der uninformierte Neuling: Projektaufsetzungsphase strukturieren

Im unerwünschten – aber manchmal unvermeidlichen – Fall, dass ein Projektleiter ein neues Projekt beginnt, ohne über ausreichende Erfahrung aus ähnlichen Projekten oder über erforderliche Daten zu verfügen, ist der Bedarf an strukturierter Tailoringunterstützung wohl am offenkundigsten. Entscheidungen müssen nicht nur vorgeschlagen und priorisiert werden, zuallererst muss ausreichend Information gesammelt werden, um gut begründete Entscheidungen zu ermöglichen. Wie oben dargestellt, ist ein auf unserem Rahmenwerk basierendes Tailoringsystem in der Lage, beiden Erfordernissen gerecht zu werden.

Angemessene Prozesse

In Abschnitt 1.1 haben wir erörtert, dass effektive Softwareentwicklung von *angemessenen* Prozessen abhängt, d. h. von Prozessen, deren Gewicht optimal auf die Anforderungen des Projekts abgestimmt ist. Mit den obigen Ausführungen haben wir gezeigt, wie werkzeuggestütztes Prozesstailoring geeignete Mittel zu diesem Zweck bietet, vorausgesetzt, es bezieht sowohl allgemeines Langfristwissen über Softwareprozesse als auch Kurzfristwissen über den Kontext des aktuellen Projekts mit ein.

These 2: Angemessene und eindeutige formale Sprache für Tailoringrichtlinien

In Kapitel 2 haben wir ein Tailoringrahmenwerk entwickelt, das uns gestattet, Tailoringrichtlinien formal in einer dreiteiligen Struktur auszudrücken, die sich aus Tailoring-Optionen, einem Tailoring-Universum, und Tailoringhypothesen zusammensetzt.

Wir können diese Struktur als formale Sprache [Wikb] betrachten, die, wie die meisten formalen Sprachen, auf einer Menge von Grundbegriffen aufbaut und die Konnektive definiert, anhand derer die Grundbegriffe zu komplexen Ausdrücken zusammengesetzt werden können.

Die von uns verwendete Sprache ist ein Spezialfall der Prädikatenlogik erster Ordnung [W+05a]: Die Grundelemente unserer Sprache sind unscharfe propositionale Variablen auf der Basis von Eigenschaften des Tailoringkontexts („die Projektdauer ist kurz") und Tailoringentscheidungen („Die Option ‚Risikomanagement' wurde ausgewählt"). Die Konnektive unserer Sprache sind die üblichen Operatoren der Prädikatenlogik [Sha05]. Tailoringhypothesen verkörpern Aussagen in unserer Sprache.

Eindeutigkeit

Wir haben eine formale Semantik zur Interpretation von Tailoringhypothesen eingeführt, indem wir Valuationsfunktionen für unscharfe propositionale Variablen und für logische Operatoren definiert haben. Die Interpretation eines beliebigen Ausdrucks in unserer Sprache hat die Form eines *unscharfen Wahrheitswert-Intervalls* und ist das Ergebnis der rekursiven Auswertung aller Bestandteile des Ausdrucks. Folglich hat jede Tailoringhypothese, die in unserer Sprache für Tailoringrichtlinien formuliert ist, eine eindeutige Interpretation.

Angemessenheit

Um zu belegen, dass unsere Sprache auch geeignet ist, Tailoringrichtlinien auszudrücken, gehen wir kurz auf alle drei oben vorgestellten Aspekte unserer Sprache ein.

Tailoring-Optionen. Unser Tailoring-Modell fußt auf der Annahme, dass Tailoring darin besteht, über jedes Element einer Menge verfügbarer Optionen eine binäre Entscheidung zu treffen – jede Option kann entweder in eine Tailoringkonfiguration aufgenommen werden, oder von ihr ausgeschlossen werden. Obwohl es in der Literatur Tailoringrichtlinien gibt, die Wahlmöglichkeiten mit mehr als nur zwei Alternativen beinhalten, wie der Ansatz von Ginsberg und Quinn (Abschnitt 5.2.1), haben wir dargelegt, dass diese auf realistische Weise auf gleichwertige binäre Optionen zurückgeführt werden können.

Tailoring-Universum. Unser Konzept von Tailoring-Universen gründet sich auf ein klar umrissenes Konzept von Metriken mit eindeutig definierten Skalentypen (siehe Tabelle 2.1 auf Seite 14), die uns gestatten, die Eigenschaften verschiedener Typen von Entitäten zu messen oder abzuschätzen. Die Unterscheidung von Entitätstypen und deren Ausprägungen erweitert den Abstraktionsgrad, unter dem Tailoringrichtlinien formuliert werden können, vor allem weil es möglich ist, mehrere Ausprägungen desselben Entitätstyps zu haben, z. B. für Teammitglieder oder Unterauftragnehmer. Zusätzliche Flexibilität wird durch die Möglichkeit eingeräumt, Messungen als Intervalle auszudrücken oder sie sogar völlig unbestimmt zu lassen, ohne damit den Tailoring-Algorithmus zu behindern.

Tailoringhypothesen. Während Ausdrücke in Prädikatenlogik es zulassen, Randbedingungen für das Tailoring in einer gängigen Art und Weise zu formulieren, bietet die Einführung von unscharfen propositionalen Variablen eine einfache aber wirksame Weise, intuitive Aussagen von Metriken abzuleiten, um so als Grundelemente von Tailoringhypothesen zu dienen. Dies gestattet eine größere Ähnlichkeit der Tailoringhypothesen zu bedingten Ausdrücken in informeller Sprache.

Diese Aspekte unseres Tailoringrahmenwerks, die so auch in unserem Tailoringbeispiel in Anhang B zur Anwendung kommen, dokumentieren, dass die zugrunde liegende formale Sprache in der Tat geeignet ist, um eindeutige und angemessene Tailoringrichtlinien auszudrücken.

These 3: Tailoringkonfigurationen nach Wertung anordnen

Wie bereits oben zusammengefasst, kann jede Tailoringhypothese in der Form eines unscharfen Wahrheitswert-Intervalls interpretiert werden. In Abschnitt 2.4.3 haben wir gezeigt, wie Bewertungen von Tailoringentscheidungen von den Erforderlichkeits- und Zulässigkeits-Hypothesen ihrer zugehörigen Tailoring-Optionen abgeleitet werden können. Anschließend haben wir in in Abschnitt 2.4.3 gezeigt, wie die Gesamtbewertung einer Tailoringkonfiguration aus den Bewertungen der sie konstituierenden Tailoringentscheidungen berechnet werden kann.

Die Bewertung jeder Tailoringentscheidung besteht aus einem Intervall innerhalb des Bereichs zwischen 0% und 100%; wir bezeichnen die Unter- und Obergrenze dieses Intervalls als die pessimistische bzw. die optimistische Bewertung. Wie in Abschnitt 2.4.4 erläutert, verwenden wir die optimistischen Bewertungen, um Tailoringkonfigurationen in Bezug auf ihre Qualität gegeneinander abzuwägen. Wir haben somit eine totale Ordnungsrelation zwischen Tailoringkonfigurationen und können These 3 als erfüllt betrachten.

These 4: Effiziente Ermittlung der optimalen Tailoringkonfiguration

Wie wir in Abschnitt 2.5 gezeigt haben, ist es uns gelungen, das Problem der Tailoring-Optimierung auf eine Suche nach kürzesten Pfaden zurückzuführen. Wir haben das Problem so umformuliert, dass es alle Voraussetzungen erfüllt, die der T*-Algorithmus – den wir vom A*-Algorithmus abgeleitet haben – fordert. Wir haben außerdem zusätzliche Optimierungen eingeführt, so eine Heuristik, die die Reihenfolge verbessert, in der Tailoringentscheidungen betrachtet werden, und indem wir den Suchraum in voneinander unabhängige Partitionen zerteilen.

In Abschnitt 4.3.3 haben wir dokumentiert, dass die Berechnung einer optimalen Tailoringkonfiguration für 41 Optionen auf einem durchschnittlichen, gängigen Einzelplatzrechner 7, 8 Sekunden dauerte. Während eine ausführliche empirische Studie über das Verhalten des Algorithmus in anderen realistischen Tailoringszenarien den Rahmen dieser Arbeit übersteigt, legen Anhaltspunkte aus ähnlichen Experimenten wie im Beispiel aus Abschnitt 4.3 nahe, dass für weniger als 150 Tailoring-Optionen die zum Auffinden der optimalen Konfiguration erforderliche Zeit annähernd proportional zur Anzahl der Tailoring-Optionen ist, dass also 150 Tailoring-Optionen in ungefähr 30 Sekunden evaluiert werden können.

Wir betrachten daher die von These 4 aufgestellte Behauptung durch die Implementierung unseres Tailoring-Algorithmus als belegt.

These 5: Wartbarkeit und Skalierbarkeit

These 5 fordert die Wartbarkeit von Tailoringrichtlinien, selbst wenn sie umfangreiche und komplexe Ausmaße annehmen. Wie wir oben erläutert haben, können wir die deskriptiven Bestandteile unseres Tailoringrahmenwerks als formale Sprache zum Ausdrücken von Tailoringrichtlinien betrachten. Die Wartbarkeit großer und komplexer Tailoringrichtlinien hängt von mindestens zwei Eigenschaften dieser formalen Sprache ab – ihrer Ausdruckskraft und ihrer Transparenz. Wir erklären nun kurz diese beiden Eigenschaften und erörtern dann ihren Einfluss auf die Wartbarkeit von Tailoringrichtlinien in unserem Tailoringrahmenwerk.

Ausdruckskraft

Unter der *Ausdruckskraft* einer Sprache verstehen wir deren Möglichkeiten, ihre Grundbegriffe zueinander in Bezug zu setzen, um neue zusammengesetzte Konzepte zu schaffen.

Wir sehen zwei grundsätzliche Arten, die Ausdruckskraft einer Sprache zu erweitern. Erstens gestattet die Einführung äquivalenter Kürzel für häufig verwendete komplexe Ausdrücke, wiederkehrende Muster knapper auszudrücken. Zweitens kann man den zugrunde liegenden semantischen Apparat erweitern und dadurch ermöglichen, Konzepte der Sprache in gänzlich neuer Weise zueinander in Bezug zu setzen.

Während der erste Ansatz neue *Formen* des Ausdrucks von Konzepten hinzufügt und somit auf die syntaktische Ebene der Sprache beschränkt ist, betrifft der zweite Ansatz die semantische Ebene der Sprache und erweitert den *Umfang* der Konzepte, die ausgedrückt werden können. Betrachen wir zum Beispiel eine Sprache, die aus den natürlichen Zahlen und der Operation der Addition besteht und die übliche Semantik der ganzzahligen Arithmetik hat. Eine syntaktische Erweiterung der Sprache wäre es, den Multiplikationsoperator einzuführen. Alle seine Vorkommen können wir anhand einer syntaktischen Transformation erklären – zum Beispiel ist 2×3 äquivalent zu $3 + 3$ oder $2 + 2 + 2$. Ein Subtraktionsoperator hingegen erweitert die Semantik unserer Sprache, da wir ihn nicht durch einen Rückgriff auf die zuvor verfügbaren Begriffe erklären können: Wir können einen Ausdruck wie $5 - 2$ nicht unter Verwendung der Additionsoperation umformulieren.

Transparenz

Mit der *Transparenz* einer Sprache bezeichnen wir den Grad, in dem die formale Interpretation von Ausdrücken in dieser Sprache mit deren intuitiver Lesart übereinstimmt. Da Tailoringrichtlinien von Menschen formuliert werden, muss die Sprache transparent sein, damit Auswertungen selbst komplexer Richtlinien mit den Erwartungen des Autors übereinstimmen.

Ausdruckskraft und Transparenz im Gleichgewicht halten

Die Ausdruckskraft und Transparenz einer Sprache hängen miteinander zusammen, und die Absicht, beide zu optimieren, leitete unsere Entwurfsentscheidungen. Wir stellten fest, dass Transparenz sich invers zu semantischer Komplexität verhält – je weniger semantische Konzepte definiert sind, umso weniger besteht die Möglichkeit unvorhergesehener Wechselwirkungen zwischen diesen Konzepten. Deshalb entschieden wir uns eine Sprache zu wählen, die zulässt, dass jede Tailoringregel auf logische Ausdrücke reduziert werden kann, die nur den Konjunktions-, Disjunktions- und Negationsoperator verwenden.

Da wir Entitätstypen und -ausprägungen unterscheiden, mussten wir auch Quantoren über mehrere Ausprägungen eines Entitätstyps einführen, aber wir beschränkten ihren Skopus auf einzelne propositionale Variablen (Abschnitt 2.4.6).

Unscharfe propositionale Variablen bringen alltagssprachliche Begriffe über Entitäten und ihre Eigenschaften ein, indem sie quantitative Messungen auf qualitative Äußerungen abbilden und somit die Notwendigkeit ausräumen, numerische Berechnungen innerhalb von Tailoringregeln zu formulieren.

Ebenfalls aus Gründen der Transparenz haben wir davon Abstand genommen, andere Sprachelemente einzuführen, wie zum Beispiel individuelle Gewichtungen für Regeln. Während solche Gewichtungen auf den ersten Blick als reizvolles Mittel erscheinen, um eine Feinabstimmung des Verhalten eines regelbasierten Systems vorzunehmen, kamen wir auf keine Umsetzung, die bei breitem Einsatz nicht zu unvorhergesehenen Nebeneffekten führen würde und somit die Wechselwirkungen zwischen Regeln verschleiern würde.

Der Bereich, in dem wir die Ausdruckskraft unserer Sprache nennenswert erweitert haben, ist die syntaktische Ebene, die es nicht nur leichter macht, Tailoringregeln auszudrücken, sondern auch ihre Transparenz erhöht, wenn zusätzliche syntaktische Konstrukte vertraute Begriffe aus der informellen Sprache widerspiegeln. Zum Beispiel liest sich der Ausdruck *wenn A dann B sonst C* leichter als $(\neg A \vee B) \wedge (A \vee C)$, obwohl sie laut der Definitionen unserer Sprache für Tailoringregeln äquivalent sind. Wie wir in Abschnitt 3.1.3 umrissen haben, haben wir mehrere solcher Operatoren eingeführt, um das Schreiben lesbarerer Tailoringregeln zu ermöglichen.

Wir haben eine weitere bedeutende Verbesserung der Ausdruckskraft mit der Einführung allgemeiner Tailoringregeln in Abschnitt 3.1.2 erreicht. Sie befreien von der Notwendigkeit, Tailoringhypothesen in Bezug auf Erforderlichkeit und Zulässigkeit einzelner Tailoring-Optionen zu formulieren und gestatten vielmehr, beliebige Zusammenhänge zwischen Tailoringentscheidungen und propositionalen Variablen auszudrücken. Diese allgemeinen Regeln werden automatisch in die eingeschränktere Form von Erforderlichkeits- und Zulässigkeits-Hypothesen sowohl für Tailoringentscheidungen als auch für propositionale Variablen transformiert. Obwohl wir deontische Logik eingeführt haben, um diese Transformation auf eine solide semantische Grundlage zu stellen, verbergen wir deontische Operatoren vor den Autoren der Regeln und umschließen stattdessen unbemerkt vom Autor jede Regel mit einem deontischen *Obligations*-Operator O.

Unter Berücksichtigung dieser Überlegungen und unserer Erfahrung mit der Modellierung von Tailoringrichtlinien für das *ReqMan*-Projekt (Abschnitt 4.3) haben wir sowohl analytische als auch empirische Argumente geliefert um unsere Behauptung zu stützen, dass ein Tailoringsystem tatsächlich die Wartbarkeit selbst großer und komplexer Tailoringrichtlinien gewährleisten kann.

These 6: Transparente Bewertungen von Tailoringkonfigurationen

In Kapitel 3.2 haben wir einen Mechanismus entworfen, der alle Tailoringentscheidungen begründet, aus denen eine Tailoringkonfiguration besteht. Diese Begründungen ermöglichen es, die Hintergründe der Bewertung einer bestimmten Konfiguration aufzuzeigen, indem alle relevanten Fakten aufgeführt werden, die zu dieser Bewertung beitragen. Diese Fakten werden in einer einfachen kausalen Hierarchie dargestellt, deren einzige Relation gleichbedeutend ist mit „*A* weil *B*," wobei die strukturelle Komplexität der zugrunde liegenden Regeln verborgen bleibt.

Wir haben in Abschnitt 3.2.2 besprochen, dass hierin eine Parallele besteht mit der Weise, in der wir bei informeller Verständigung Begründungen vorbringen. Die landläufige Art, eine Behauptung zu untermauern besteht darin, Tatsachen zu nennen, die die Behauptung stützen – in der stillschweigenden Annahme, dass eine Kausalbeziehung besteht, deren Beschaffenheit nicht näher ausgeführt wird. Wir sagen „zieh' deinen Mantel an, denn *es regnet*," und nicht „zieh' deinen Mantel an, denn *er wird verhindern, dass du im Regen nass wirst*."

Das Ergebnis dieses Ansatzes ist, dass der Benutzer des Tailoringsystems die Begründungen schneller durchgehen kann, weil ihm die Begründungen in einer gewohnten Form dargeboten werden, und weil er keine komplexen Strukturen analysieren muss, um die Begründungen zu verstehen: Stattdessen wird alle Information in Form von einfachen, unabhängigen Aussagen bereitgestellt.

Der Einsatz der Logik unscharfer Wahrheitsintervalle verschafft in diesem Zusammenhang mehrere Vorteile. Unscharfe Variablen sind eng verwandt mit ungefähren Aussagen, die oft im menschlichen logischen Denken anzutreffen sind, und gestatten die Konstruktion intuitiver informeller Aussagen über den Tailoringkontext. Weiterhin drücken Intervalle unscharfer Wahrheitswerte den Grad sowohl des Zutreffens wie auch der Sicherheit einer Aussage aus. Wahrheitswerte um die 50% besagen, dass eine Aussage weder gänzlich wahr noch gänzlich widerlegbar ist, während Wahrheitswerte nahe eines der beiden Enden der Skala von 0% bis 100% eine Aussage ausdrücklich verwerfen oder anerkennen. Gleichermaßen gilt, dass mit der zunehmenden Breite eines Wahrheitsintervalls die Sicherheit über die damit verbundene Aussage sinkt.

Mit dem Rückgriff auf eine – wie in Abschnitt 3.2.5 vorgeschlagene – grafische Darstellung von Begründungen und den darin enthaltenen bewerteten Aussagen können Zutreffen und Sicherheit solcher Aussagen visuell und intuitiv erkannt werden.

Anhang C bietet ein anschauliches Beispiel einer begründeten Tailoringkonfiguration. Diese und die obigen Überlegungen lassen uns zu dem Schluss kommen, dass es uns gelungen ist eine Methode zu entwickeln, anhand derer die Bewertung einer Tailoringkonfiguration auf transparente Weise möglich ist, ohne Wissen über die zugrunde liegenden Regeln vorauszusetzen.

Weiterführende Fragestellungen

Wie im Fall aller wissenschaftlichen Problemstellungen haben auch wir, während wir mit dem Lösen der uns selbst gesteckten Herausforderungen befasst waren, neue Problemstellungen aufgedeckt, in denen wir viel versprechende Themen für zukünftige Forschungen sehen. Wir sind auf drei zentrale Fragen gestoßen, die wir für die weiteren Untersuchung unseres Tailoringrahmenwerks als lohnend erachten: Entwurfskriterien für Tailoringrichtlinien, eine zweckmäßige Benutzerschnittstelle, und weitere Anwendungsbereiche.

Entwurfskriterien für Tailoringrichtlinien. Wir haben eine Sprache zur Formulierung von Tailoringrichtlinien eingeführt. Wie bei jeder neuen Sprache zählt zu den grundlegenden Fragen, auf welche Weise sie am besten eingesetzt werden kann. Wir haben Syntax und Semantik unserer Sprache abgedeckt. Nicht abgedeckt haben wir Stilfragen. Welche stilistischen Konventionen und Entwurfskriterien für Tailoringrichtlinien gestatten am besten, die Möglichkeiten der Sprache auszuschöpfen? Wie können Tailoringregeln formuliert und gegliedert werden, um deren Verständlichkeit, Wartbarkeit, Exaktheit und Wiederverwendungsmöglichkeiten zu maximieren? Weitere Erfahrungen mit praxisbezogenen Anwendungen unseres Tailoringrahmenwerks werden das Material und die Erfahrung einbringen, die zum Beantworten dieser Fragen erforderlich sind.

Zweckmäßige Benutzerschnittstelle. Wie könnte eine zweckmäßige Benutzerschnittstelle für unser Tailoringsystem aussehen? Das hängt davon ab, wie und unter welchen Umständen unser Tailoringrahmenwerk eingesetzt werden soll. In Abschnitt 3 haben wir zwei Hauptarten von Anwendern festgestellt – den Prozessmodellierer und den Prozesstailorer. In Abschnitt 4.3.5 haben wir bereits erörtert, dass beide Anwenderarten von einem geeigneten Softwarewerkzeug profitieren würden, das den Zugriff auf unser Rahmenwerk gestattet, und wir haben grundsätzliche Eigenschaften eines solchen Werkzeugs umrissen.

In Abschnitt 5.2.3 haben wir unseren regelbasierten Ansatz mit fallbasierten Ansätzen verglichen. Ein Softwarewerkzeug könnte unseren regelbasierten Ansatz um die Vorteile eines fallbasierten Ansatzes erweitern, indem es eine Datenbank zuvor durch Tailoring angepasster Prozesse unterhält, und könnte mit einem geeigneten Vergleichsalgorithmus bereits angepasste Prozesse aus ähnlichen, in der Vergangenheit erfolgreichen Projekten abrufen. Da ein Prozesstailorer in den meisten Fällen eine vom Tailoringsystem empfohlene Tailoringkonfiguration anpassen und feinjustieren wird (siehe Abschnitt 4.3.3), könnte der Vergleich seiner Entscheidungen mit Tailoringkonfigurationen aus früheren Projekten eine sinnvolle Erweiterung sein.

Ein anderes Grundprinzip des fallbasierten Ansatzes ist die kontinuierliche Bereicherung der Datenbasis um Erfahrungen aus neuen Projekten. Um die Einbeziehung solcher neuer

Erfahrungen in ein regelbasiertes Tailoringsystem zu gestatten, ist es unverzichtbar, dass Tailoringregeln leicht zu verwalten und verständlich sind, damit eine größere Expertengemeinschaft motiviert und in der Lage ist, ihre Erfahrungen regelmäßig in der Regelbasis aufzuzeichnen.

Weitere Anwendungsbereiche. Unsere Arbeit bezieht ihre Anregung aus praktischen Problemen der Erstellung angemessener Prozesse durch Tailoring im Kontext der Softwareentwicklung, und wir haben einen Ansatz aufgestellt, der zu diesem Zweck besonders geeignet ist.

Unser Tailoring-Kernrahmenwerk basiert jedoch auf einem sehr allgemeinen Tailoringbegriff, und trifft keine Annahmen über Eigenheiten der Softwareentwicklung, sogar nicht einmal von Prozessen: Solange ein Tailoringproblem in irgendeinem anderen Themenfeld in seiner Struktur den Definitionen unseres Tailoringrahmenwerks entspricht – das Treffen einer optimale Auswahl aus verfügbaren Optionen mit Hinblick auf eine Kontextbeschreibung – kann unser Tailoringrahmenwerk darauf ebenso gut angewandt werden.

Folglich könnte unser Ansatz auch auf andere Arten von Prozessen angewandt werden, wie beispielsweise auf die gesamte Bandbreite der vom CMMI [Kas04] abgedeckten Prozesse für System-Engineering und -Management, auf die in der *IT Infrastructure Library* (ITIL) [ITI] aufgestellten *best-practice*-Prozesse für IT-Management, oder überhaupt auf jeden Prozess, der formal dokumentiert ist.

Der Anwendungsbereich könnte sogar noch weiter vergrößert werden, um weiteren Problemstellungen wie den folgenden gerecht zu werden:

- Anschaffung konfigurierbarer Geräte für einen bestimmten Verwendungszweck, wobei Tailoring-Optionen für optionale Komponenten stehen

- standardisierte Produkte und Dienstleistungen konfigurieren, wie zum Beispiel Autos oder Mobiltelefonverträge

- individuelle Studienpläne aus Universitätsvorlesungen zusammenstellen, und dabei sicher stellen, dass die Kursauswahl den Mindestanforderungen des angestrebten Abschlusses entspricht

- Trainingspläne für die Mitglieder eines Fitnessstudios zusammenstellen unter Beachtung von Alter, Geschlecht, allgemeiner Gesundheit, und Trainingszielen

Ausblick

Wir glauben, dass viele Leute allergisch auf Prozesse reagieren, weil in vielen Fällen Prozesse nicht an die besonderen Erfordernisse eines Projekts angepasst sind. Infolgedessen wird

Agilität oft missverstanden als das Fehlen eines eindeutig definierten Prozesses im Gegensatz zu der Rigidität, die von einem definierten Prozess erzwungen wird. Die eigentliche Entscheidung liegt hingegen nicht darin, einen Prozess zu *haben* oder ihn *nicht zu haben*, sondern die Entscheidung besteht darin, ob man sich des Prozesses *nicht bewusst* ist oder ob man ihn *bewusst* gestaltet. Softwaregestütztes Tailoring ist ein effektives Werkzeug, um ein Prozessmodell den besonderen Erfordernissen eines Projekts anzupassen. Wir gehen davon aus, dass es das Tailoring nicht nur vereinfachen wird, sondern dass es auch zur weiteren Verbreitung des Tailoring beitragen wird. Tailoring ist unumgänglich, um die scharfe Trennung zwischen Agilität und Rigidität zu überwinden, denn es zielt darauf ab, immer denjenigen Prozess zu erhalten, der so agil wie möglich, aber so rigide wie nötig ist. Wir hoffen daher, dass unsere Arbeit dazu beiträgt, die Rolle des Tailoring in der Praxis der Softwareentwicklung zu stärken.

Bibliography

[ADH⁺01] ALTHOFF, Klaus-Dieter ; DECKER, Björn ; HARTKOPF, Susanne ; JED-LITSCHKA, Andreas ; NICK, Markus ; RECH, Jörg: Experience Management: The Fraunhofer IESE Experience Factory. In PERNER, P. (Ed.): *Industrial Conference Data Mining*. Leipzig, Germany, July 2001

[AgA] *The Agile Alliance*. Web site. http://www.agilealliance.com

[AgM] *The Agile Manifesto*. Web site. http://www.agilemanifesto.org

[APM] Appropriate Process Group: *Appropriate Process Movement*. Web site. http://www.aptprocess.com

[Bec01] BECK, Kent: *Extreme Programming Explained: Embrace Change*. Addison Wesley, 2001

[BPM05] Object Management Group/Business Process Management Initiative: *BPMN Information Home*. Internet Portal. http://www.bpmn.org/. 2005

[BR87] BASILI, V. R. ; ROMBACH, H. D.: Tailoring the software process to project goals and environments. In *Proceedings of the 9th international conference on Software Engineering*. Monterey, California, United States : IEEE Computer Society Press, 1987, p. 345–357

[BR88] BASILI, V. R. ; ROMBACH, H. D.: The TAME Project: Towards improvement-oriented software environments. In *IEEE Transactions on Software Engineering* 14 (1988), June, No. 6, p. 758–773

[Bre03] BRENNAN, Andrew: Necessary and Sufficient Conditions. Fall 2003. http://plato.stanford.edu/archives/fall2003/entries/necessary-sufficient/. In ZALTA, Edward N. (Ed.): *The Stanford Encyclopedia of Philosophy*. Stanford University

[BT03] BOEHM, Barry ; TURNER, Richard: Using Risk to Balance Agile and Plan-Driven Methods. In *IEEE Computer* Vol. 36. IEEE Computer Society, June 2003

[CFF94]   CONRADI, R. ; FERNSTRÖM, C. ; FUGGETTA, A.: Concepts for Evolving Software Process. In FINKELSTEIN, A. (Ed.) ; KRAMER, J. (Ed.) ; NUSEIBEH, B. (Ed.): *Software Process Modelling and Technology*. Research Studies Press. Wiley and Sons, 1994

[CG98]   CUGOLA, Gianpaolo ; GHEZZI, Carlo: Software Processes: a Retrospective and a Path to the Future. In *Software Process: Improvement and Practice* 4 (1998), No. 3, 101-123. http://citeseer.ist.psu.edu/cugola98software.html

[Coc02]   COCKBURN, Alistair: *Agile Software Development*. Addison Wesley, 2002

[Cop01]   COPELAND, Lee: Extreme programming in the U.K.   In *Computerworld* (2001), April. http://www.itworld.com/AppDev/1254/CWD010409STO59388

[CVW+01]   CASS, A. ; VÖLCKER, C. ; WINZER, L. ; CARRANZA, J. M. ; DORLING, A.: SPiCE for SPACE: A Process Assessment and Improvement Method for Space Software Development. In *ESA Bulletin*. European Space Agency, August 2001 (107)

[ED96]   EBERT, Christof (Ed.) ; DUMKE, Reiner (Ed.):   *Software-Metriken in der Praxis*. Springer, 1996

[FH93]   FEILER, Peter H. ; HUMPHREY, Watts S.: Software Process Development and Enactment: Concepts and Definitions. In *ICSP2*, 1993

[FN86]   FORSTYH, Richard ; NAYLOR, Chris: *The Hitch-Hiker's Guide to Artificial Intelligence*. London : Chapman and Hall, 1986

[Fow99]   FOWLER, Martin: *Refactoring*. Addison-Wesley Professional, 1999

[FRG04]   Federal Republic of Germany, Ministry of the Interior: *V-Modell® XT*. Web site. http://www.v-modell-xt.de. Version 1.1.0, 2004

[Gar05]   GARSON, James: Modal Logic.   Summer 2005.   http://plato.stanford.edu/archives/sum2005/entries/logic-modal/.   In ZALTA, Edward N. (Ed.): *The Stanford Encyclopedia of Philosophy*. Stanford University

[Gla01]   GLAZER, Hillel: Dispelling the Process Myth: Having a Process Does Not Mean Sacrificing Agility or Creativity. In *CrossTalk* (2001), November

[GQ94]   GINSBERG, Mark P. ; QUINN, Lauren H.: Process Tailoring and the Software Capability Maturity Model / Carnegie Mellon Software Engineering Institute. 1994 (CMU/SEI-94-TR-024). – technical report

[HHMS04]   HINDEL, Bernd ; HÖRMANN, Klaus ; MÜLLER, Markus ; SCHMIED, Jürgen: *Basiswissen Software-Projektmanagement*. dpunkt.verlag, 2004

[Hin96]   HINDEL, Bernd: Qualität ist meßbar: Software-Metriken. In *Design &* *Elektronik* (1996), November, p. 50–55

[HIS]   *Herstellerinitiative Software.* Web site. `http://www.automotive-his.de`

[HMVV04]   HINDEL, Bernd ; MEIER, Erich ; VLASAN, Adriana ; VERSTEEGEN, Gerhard: *Prozessübergreifendes Projektmanagement.* Springer, 2004

[Hop03]   HOPEN, Sigurd: *IBM Rational Unified Process—A scalable software development process?* Commercial presentation, 2003

[HRS04]   HRUSCHKA, Peter ; RUPP, Christine ; STARKE, Gernot: *Agility kompakt.* Spektrum Akademischer Verlag, 2004

[HV03]   HINDEL, Bernd ; VLASAN, Adriana: Reif für Dokumentation – Wie agil dürfen Dokumente sein? In *Automatisierungstechnische Praxis* (2003), November

[IAB04]   IABG: *V-Model® XT.* Press release. `http://www.iabg.de/presse/aktuelles/mitteilungen/200409_V-Model_XT_en.php`. September 2004

[IBM]   IBM Software: *The Rational Unified Process.* Web page. `http://www.ibm.com/software/awdtools/rup/`

[IEC]   The 61508 Association: *What is the IEC 61508?* Web page. `http://www.61508.org/61508.htm`

[IEC98]   Functional safety of electrical/electronic/programmable electronic safety-related systems – Part 3: Software requirements / IEC. 1998 (TR 61508-3:1998). – technical report

[IEC05]   Functional safety of electrical/electronic/programmable electronic safety-related systems – Part 0: Functional safety and IEC 61508 / IEC. 2005 (TR 61508-0:2005). – technical report

[Inf]   *Infosys.* Web site. `http://www.infosys.com`

[ISO95]   Information Technology – Software Life Cycle Processes / ISO/IEC. 1995 (ISO/IEC 12207). – technical report

[ISO98a]   Information technology – Software process assessment / ISO/IEC. 1998 (ISO/IEC TR 15504-5). – technical report

[ISO98b]   Information technology – Software process assessment – Part 5: An assessment model and indicator guidance / ISO/IEC. 1998 (ISO/IEC TR 15504-5). – technical report

[ISO05]   ISO/IEC: *ISO/IEC 15504*. Standards document. http://www.iso.org. 2003–2005

[ITI]   *IT Infrastructure Library (ITIL)*. Web site. http://www.itil.co.uk

[J⁺92]   JACOBSON, Ivar et al.: *Object-Oriented Software Engineering: A Use-Case-Driven Approach*. Reading, MA : Addison-Wesley, 1992

[JAH00]   JEFFRIES, Ronald E. ; ANDERSON, Ann ; HENDRICKSON, Chet: *Extreme Programming Installed*. Addison-Wesley Professional, 2000

[Jal99]   JALOTE, Pankaj: *Cmm in Practice*. Addison-Wesley, 1999

[JBR98]   JACOBSON, Ivar ; BOOCH, Grady ; RUMBAUGH, James: *The Unified Software Development Process*. Reading, MA : Addison Wesley Longman, 1998

[Kas04]   KASSE, Tim: *Practical Insight into CMMI*. Artech House, 2004 (computing library)

[Kee04]   KEENAN, Frank: Agile Process Tailoring and probLem analYsis (APTLY). In *Proceedings of the 26th International Conference on Software Engineering* University of Ulster at Coleraine, IEEE Computer Society, 2004, p. 45–47

[Kol93]   KOLODNER, J. L.: *Case-Based Reasoning*. Morgan Kaufmann Publishers, 1993

[Kru00]   KRUCHTEN, Philippe: *The Rational Unified Process — An Introduction*. Reading, MA : Addison-Wesley, 2000

[Lau02]   LAUESEN, Soren: *Software Requirements. Styles and Techniques*. 1st edition. Addison-Wesley, 2002

[Lon93]   LONCHAMP, J.: A Structured Conceptual and Terminological Framework for Software Process Engineering. In *ICSP2*. Berlin, February 1993, p. 41–53

[LR93]   LOTT, Christopher M. ; ROMBACH, H. D.: Measurement-Based Guidance of Software Projects Using Explicit Project Plans. In *Information and Software Technology* Volume 35 (1993), June, No. 6

[LR98]   LIENERT, Gustav A. ; RAATZ, Ulrich: *Testaufbau und Testanalyse*. 6. Auflage. BeltzPVU, 1998

[Miš05]   MIŠIČ, Vojislav B.: Perceptions of Extreme Programming: A Pilot Study / Department of Computer Science, University of Manitoba. 2005. http://www.cs.umanitoba.ca/~vmisic/pubs/tr0503.pdf (05/03). – technical report. – online resource

[MT77]   MOSTELLER, Frederick ; TUKEY, John: *Data analysis and regression*. Addison-Wesley, 1977

[O⁺05] OLSSON, Thomas et al.: *RE-Wissen*. Web portal. `http://www.re-wissen.de`. 2005

[ODKE05] OLSSON, Thomas ; DOERR, Joerg ; KOENIG, Tom ; EHRESMANN, Michael: A Flexible and Pragmatic Requirements Engineering Framework for SME. In *Proceedings of SREP '05*. Paris, 2005

[ØH95] ØHRSTRØM, Peter ; HASLE, Per F.: *Temporal Logic. From Ancient Ideas to Artificial Intelligence*. Dordrecht : Kluwer Academic Publishers, 1995

[Old03] OLDFIELD, Paul: *Appropriate Process Approaches*. Whitepaper, 2003

[PK] *project>kit*. Web page. `http://www.methodpark.de/products/projectkit/`

[RD00] REITER, Ehud ; DALE, Robert: *Bulding Natural Language Generation Systems*. Cambridge, UK : Cambridge University Press, 2000

[Ric00] RICHTER, Michael: Fallbasiertes Schließen. In GÖRZ, G. (Ed.) ; ROLLINGER, C.-R. (Ed.) ; SCHNEEBERGER, J. (Ed.): *Handbuch der Künstlichen Intelligenz*. Oldenbourg, 2000, Chapter 11, p. 407–430

[RM] *ReqMan*. Web site. `http://www.reqman.de`

[Roy70] ROYCE, W. W.: Managing the Development of Large Software Systems: Concepts and Techniques. In *Proceedings of WesCon*, 1970

[SB02] SCHWABER, Ken ; BEEDLE, Mike: *Agile Software Development with Scrum*. Pentice Hall, 2002

[Sch99] SCHEER, August-Wilhelm: *ARIS – Business Process Frameworks*. 3rd edition. Springer, 1999

[SE43] SAINT-EXUPÉRY, Antoine de: *Le Petit Prince*. Harcourt Brace, 1943

[SEI] *Carnegie Mellon University Software Engineering Institute (SEI)*. Web site. `http://www.sei.cmu.edu`

[SEI02a] Capability Maturity Model® Integration / Carnegie Mellon Software Engineering Institute. August 2002. `http://www.sei.cmu.edu/cmmi/models/models.html`. – technical report. – online resource

[SEI02b] Capability Maturity Model® Integration (CMMI[SM]), Version 1.1, Continuous Representation / Carnegie Mellon Software Engineering Institute. August 2002. `http://www.sei.cmu.edu/publications/documents/02.reports/02tr011.html`. – technical report. – online resource

[Sha05]   SHAPIRO, Stewart: Classical Logic. Fall 2005. http://plato.stanford.
            edu/archives/fall2005/entries/logic-classical/. In ZALTA,
            Edward N. (Ed.): *The Stanford Encyclopedia of Philosophy*. Stanford Uni-
            versity

[SO01]   SINDRE, G. ; OPDAHL, A.: Templates for Misuse Case Description. In *7th In-
            ternational Workshop on Requirements Engineering (REFSQ'2001)*. Switzer-
            land, 4–5 June 2001

[SR03]   STEPHENS, Matt ; ROSENBERG, Doug: *Extreme Programming Refactored:
            The Case Against XP*. Apress, 2003

[Ste46]   STEVENS, Stanley S.: On the theory of scales of measurement. In *Science*
            (1946), No. 103, p. 677–680

[Sun]   Sun Developer Network: *Java Technology*. Web site. http://java.sun.
            com/

[VW93]   VELLEMAN, Paul F. ; WILKINSON, Leland: Nominal, Ordinal, Interval, and
            Ratio Typologies are Misleading. In *The American Statistician* 47 (1993),
            No. 1, p. 65–72

[W+05a]   WEISSTEIN, Eric W. et al.: First-Order Logic. 2005. http://mathworld.
            wolfram.com/First-OrderLogic.html. In *MathWorld*. Wolfram Re-
            search, Inc.

[W+05b]   WEISSTEIN, Eric W. et al.: Propositional Calculus. 2005. http://
            mathworld.wolfram.com/PropositionalCalculus.html. In *Math-
            World*. Wolfram Research, Inc.

[Wika]   Wikipedia: *A* search algorithm*. Web page. http://en.wikipedia.
            org/wiki/A-star_search_algorithm

[Wikb]   Wikipedia: *Formal Language*. Web page. http://en.wikipedia.org/
            wiki/Formal_language

[Wikc]   Wikipedia: *KSLOC*. Web page. http://en.wikipedia.org/wiki/
            KSLOC

[Wikd]   Wikipedia: *Process*. Web page. http://en.wikipedia.org/wiki/
            Process

[Xu05]   XU, Peng: Knowledge Support in Process Tailoring. In *Proceedings of the
            38th Hawaii International Conference on System Sciences*, 2005

[Zad65]   ZADEH, Lotfi A.: Fuzzy Sets. In *Information and Control* 8 (1965), p. 338–
            353

Index

A*algorithm, 43
APTLY, 114
artificial intelligence, 103
Automotive SPiCE, 107
axiom
    of justification, 71

bias function, 30

case-based reasoning, 113
closed world assumption, 66
CMM
    tailoring table, *see* tailoring
CMMI, 104–106
    model tailoring, *see* tailoring
    practice, 104
    process area, 104
    process tailoring, *see* tailoring
committee, 35
Crystal Methodologies, 113

definiteness margin, 75
deontic logic, *see* logic
development case (RUP), 107

entity, 12
ex falso quodlibet, 67
Experience Factory, 113

feasible, 26
formula
    fuzzy, *see* fuzzy
functional safety, 108
fuzzy
    formula, 20
      atomic, 52

interval logic, 22
logic, 19
propositional variable, 19
truth interval, 22
truth value, 19
valuation function, 21

HTML
    generation, 90

IEC
    15504, *see* SPiCE
    61508, 108
Infosys, 111
ISO/IEC, *see* IEC

Java, 89
justification, 70
    graphical representation of, 85
    hierarchy, 70
      construction, 76
    of decision ratings, 77–80
    of proposition valuation, 72–77

language
    formal, 4, 129
    natural
      generation, 87
      misinterpretations of, 57
      statements in, 71
LaTeX
    generation, 90
life cycle, 10
    model
      CMMI, 105
logic

223

www.ingramcontent.com/pod-product-compliance
Lightning Source LLC
LaVergne TN
LVHW062313060326
832902LV00013B/2196